Scent of Sunshine Fields

Uncovering woodland ecosystems and vibrant plant relationships

Rían Chatham

Contents

1 Chapter 1: Woodland Introduction

Woodlands, at their heart, are more than just collections of trees; they are intricate tapestries woven from the threads of life. They represent a fascinating middle ground, a space distinct from the dense, shaded realm of a true forest and the wide-open expanse of a grassland. Understanding their unique characteristics is the first step in appreciating the complex beauty they hold.

The defining feature of a woodland is its openness. Unlike forests where the canopy forms a near-continuous ceiling, woodlands have a more scattered distribution of trees. Sunlight penetrates more readily, reaching the ground and fostering a rich understory of grasses, shrubs, and wildflowers. This increased light availability is a crucial factor that shapes the entire ecosystem, influencing the types of plants and animals that can thrive there. Soil conditions also play a massive part in woodland creation and structure. Well-draining soils are a plus, as saturated or too acidic ones are not ideal for woodlands. Geology is another critical factor that plays a role in woodland structure.

Think of it as a mosaic, where patches of sunlight and shade create a dynamic environment. This mosaic effect influences everything from the temperature and humidity at ground level to the types of insects that can survive. This also gives rise to microclimates, which can result in some plants thriving here and others struggling to adapt and flourish, especially for plants requiring high levels of sunlight for survival. These conditions are highly influential in how the plants eventually evolve and adapt to these conditions.

However, the term "woodland" isn't as straightforward as one might think. Its precise meaning can vary depending on the region and the context in which it's used. In some areas, it might refer to any area dominated by trees, regardless of the density. In other regions, a stricter definition is applied, emphasizing the open canopy and the presence of a well-developed understory. This ambiguity highlights the importance of considering the specific ecological context when studying or describing a woodland. A woodland in the Mediterranean will look completely different to one in Scandinavia.

Beyond the physical structure and the light regime, water is important. Areas that have regular and predictable rainfall are better suited than those regions with unpredictable weather conditions. Areas with lower rainfall will affect the diversity of plants and animals in the region.

Now, let's talk about biodiversity. It's the lifeblood of any ecosystem, and woodlands are no exception. The greater the diversity of species, the more resilient the ecosystem is to disturbances like disease, climate change, and even human interference. Woodlands are home to an impressive array of organisms, from the tallest trees to the tiniest microbes in the soil.

Consider the plant life, for example. The dominant trees, of course, form the backbone of the woodland. Oak, birch, maple, and pine are common species, each with its own specific requirements and adaptations. Beneath the trees, a layer of shrubs, such as hazel, hawthorn, and dogwood, provides shelter and food for many animals. Then, at ground level, a carpet of grasses, ferns, mosses, and wildflowers creates a vibrant tapestry. The presence of each plant affects the other.

But plants are only one piece of the puzzle. Woodlands also teem with animal life. Insects are abundant, playing crucial roles in pollination, decomposition, and nutrient cycling. Birds flit through the branches, singing their melodies and feasting on insects and seeds. Mammals, from tiny voles and mice to larger deer and foxes, roam the understory, each contributing to the complex food web. From the smallest ant to the largest bear, there is a specific role for each of these to play.

The interconnections between these organisms are what truly make a woodland ecosystem special. A pollinator, for example, feeds on the nectar of a flower, carrying pollen from one plant to another and enabling reproduction. A herbivore grazes on plants, providing nutrients for decomposers when it dies. Predators control populations, preventing any one species from becoming too dominant. Every organism has a role to play, and the health of the ecosystem depends on the proper functioning of all these interactions.

Take, for instance, the relationship between oak trees and acorns. Oak trees produce acorns, which are a vital food source for many animals, including squirrels, deer, and wild turkeys. These animals, in turn, help to disperse the acorns, spreading them to new locations where they can germinate and grow into new oak trees. The acorns that are not eaten contribute to the soil and eventually provide nutrients for new plants to grow. This seemingly simple interaction highlights the intricate web of dependencies that characterize a woodland ecosystem.

We cannot disregard the crucial role of fungi and microorganisms. These often-unseen organisms play a vital role in decomposition, breaking down dead organic matter and releasing nutrients back into the soil. They also form symbiotic relationships with plant roots, helping them to absorb water and nutrients more efficiently. These microbial communities are essential for maintaining the health and productivity of the woodland ecosystem. This contributes to the continuous cycle of growth, death and rebirth.

Ultimately, the biodiversity of a woodland is what gives it its resilience. A diverse ecosystem is better able to withstand disturbances and adapt to changing conditions. For example, if a particular plant species is susceptible to a disease, the presence of other plant species can help to prevent the disease from spreading and decimating the entire plant community. Similarly, a diverse animal community can provide a buffer against population fluctuations, ensuring that the ecosystem remains stable even in the face of environmental changes.

Exploring and understanding these ecosystems requires a deep respect for nature. As we delve deeper into the specific aspects of woodland ecology in the following chapters, let us remember the importance of preserving these invaluable ecosystems for future generations. The more you explore, the more the woodland will reveal.

1.1 What is a Woodland Ecosystem?

Imagine standing at the edge of a world painted in shades of green and brown, sunlight dappling through a canopy of leaves. This is a woodland, a place where trees dominate, but not with the dense, overshadowing presence of a forest. Instead, a woodland breathes, its trees more widely spaced, allowing sunlight to reach the ground, nurturing a diverse understory of shrubs, grasses, and wildflowers. To truly understand this environment, we must delve into its defining characteristics and appreciate the complex interactions that make it a unique and vital ecosystem.

A woodland isn't just a collection of trees; it's a complex web of life, a dynamic community where plants, animals, fungi, and microorganisms all play essential roles. The key difference between a woodland and a forest lies primarily in the density of the tree cover. Forests typically have a closed canopy, meaning the treetops touch or overlap, creating deep shade below. Woodlands, on the other hand, have a more open canopy, allowing significantly more sunlight to penetrate to the ground. This light availability is crucial, shaping the type of vegetation that can thrive beneath the trees. This difference in light penetration also allows more plants to grow below the canopy. This light availability also creates many different types of vegetation and trees to prosper in this biome.

Several factors contribute to the formation and maintenance of woodland ecosystems. Climate plays a vital role. Woodlands are typically found in regions with moderate rainfall and temperatures, although they can also occur in drier areas with specific adaptations. Soil type is another key determinant, influencing the types of trees and other plants that can grow. Woodlands are also shaped by natural disturbances such as fire, grazing, and windthrow, which can create gaps in the canopy, promoting regeneration and biodiversity. The human impact, including deforestation, agriculture, and introduction of invasive species, has significantly changed and continues to affect many woodland ecosystems.

One of the most remarkable aspects of woodlands is their biodiversity. The open canopy and varied light conditions support a wide array of plant life, from towering trees like oak, birch, and maple, to smaller trees like dogwood and hawthorn. Shrubs such as hazel, elderberry, and honeysuckle add to the complexity of the understory. Ground cover plants, including grasses, ferns, and wildflowers like bluebells and primroses, create a tapestry of color and texture. The plant diversity, in turn, supports a rich fauna. Insects, birds, mammals, reptiles, and amphibians find food, shelter, and breeding grounds within the woodland. From the tiniest beetle to the largest deer, each creature contributes to the overall health and functioning of the ecosystem.

To appreciate the significance of biodiversity, consider the interconnectedness of species within a woodland. Plants provide food and habitat for insects, which, in turn, are eaten by birds. Birds help disperse seeds, contributing to plant regeneration. Mammals like squirrels and voles consume seeds and nuts, influencing plant distribution. Fungi decompose dead organic matter, releasing nutrients back into the soil. Earthworms aerate the soil, improving drainage and nutrient availability. These interactions, and countless others, create a delicate balance that sustains the entire ecosystem. Losing even a single species can have cascading effects, disrupting the food web and potentially leading to further losses of biodiversity.

The woodland also provides essential ecosystem services that benefit humans. Woodlands help regulate water cycles, preventing soil erosion and reducing the risk of flooding. The trees absorb carbon dioxide from the atmosphere, helping to mitigate climate change. Woodlands purify air and water, providing cleaner resources for human consumption. They offer recreational opportunities, such as hiking, birdwatching, and nature photography, promoting human well-being. Moreover, woodlands have cultural and aesthetic value, inspiring art, literature, and spiritual connection. For example, some people may not realize the amount of recreation that woodland offers, and that it can have a huge positive impact on the mental health of those who engage in those activities.

Understanding the structure of a woodland ecosystem is crucial to appreciating its complexity. The canopy, composed of the tallest trees, intercepts much of the sunlight and rainfall. The understory, consisting of smaller trees and shrubs, provides habitat for many animals. The ground cover layer, including grasses, ferns, and wildflowers, plays a vital role in nutrient cycling and soil stabilization. Each layer has its unique microclimate, influencing the distribution of plants and animals. For example, the canopy trees are well-adapted to withstand strong winds and intense sunlight, while the ground cover plants thrive in the shade and moisture.

Woodlands are also incredibly dynamic environments, constantly changing in response to seasonal variations, disturbances, and long-term ecological processes. In spring, wildflowers bloom, attracting pollinators and providing early food sources for insects. In summer, the canopy is in full leaf, providing shade and cooling the understory. In autumn, leaves change color, creating a spectacular display before falling to the ground and enriching the soil. In winter, the woodland may appear dormant, but life continues beneath the surface as seeds germinate and animals seek shelter.

However, woodlands worldwide are facing increasing threats. Habitat loss due to deforestation and urbanization is a major concern. Climate change is altering temperature and rainfall patterns, affecting the distribution and health of trees and other plants. Invasive species are outcompeting native plants and disrupting ecological processes. Pollution from industrial and agricultural sources is contaminating soil and water, harming wildlife. Overgrazing by livestock can damage the understory and prevent tree regeneration. In order to truly preserve the beauty and richness of the woodland, humans must come together and develop ways to maintain the land.

Protecting and restoring woodland ecosystems requires a multifaceted approach. Conservation efforts include preserving existing woodlands through protected areas, such as national parks and nature reserves. Reforestation and afforestation projects aim to reestablish woodlands on degraded lands. Sustainable forestry practices promote responsible harvesting and management of timber resources. Control of invasive species is crucial for maintaining biodiversity. Reducing pollution and mitigating climate change are essential for the long-term health of woodlands. Education and public awareness campaigns can inspire people to appreciate and protect these valuable ecosystems.

I remember a time when I was walking through a local woodland area and stumbled upon a section where a group of volunteers were actively planting native trees and shrubs. The enthusiasm and care that these individuals displayed served as a powerful reminder of the importance of collective action. By embracing responsible stewardship and promoting sustainable practices, we can ensure that future generations will have the opportunity to experience the magic and wonder of woodland ecosystems. The scent of damp earth, the rustle of leaves in the breeze, the vibrant colors of wildflowers – these are the treasures that we must strive to protect.

1.2 Exploring Biodiversity

Imagine stepping into a world teeming with life, a vibrant tapestry woven with countless threads of interconnected organisms. This, in essence, is biodiversity, and its importance in woodland ecosystems cannot be overstated. It's far more than just a checklist of species; it represents the complex web of interactions that sustains the entire system. This intricate network ensures stability, resilience, and the continued provision of essential ecosystem services.

Biodiversity within a woodland is apparent in every layer, from the towering canopy trees to the smallest soil microorganisms. A diverse canopy, comprised of varying tree species and age classes, creates a mosaic of light and shade, influencing temperature and humidity levels throughout the woodland. This heterogeneity fosters a wider range of habitats, supporting a greater variety of understory plants, insects, and animals. Think of it as an architectural masterpiece, designed by nature with each component playing a vital role. A homogenous canopy, dominated by a single species, offers limited niches, restricting the overall biodiversity of the system. It's analogous to living in a house with only one room – functional, perhaps, but lacking in variety and adaptability.

The understory, often overlooked, is a hotbed of biodiversity. Shrubs, saplings, and herbaceous plants compete for resources, creating a dynamic environment that supports a diverse community of herbivores and pollinators. Different plant species offer different nutritional values and physical structures, catering to a range of insect larvae, browsing mammals, and nesting birds. For example, a hawthorn hedge provides shelter for nesting birds and berries for winter sustenance, while a patch of brambles offers food and refuge for insects and small mammals. Removing this layer significantly reduces the structural complexity of the woodland, impacting the species that rely on it.

Consider the ground layer, a world hidden beneath our feet. Here, mosses, lichens, and fungi thrive, playing crucial roles in nutrient cycling and decomposition. Fungi, particularly mycorrhizal fungi, form symbiotic relationships with tree roots, enhancing nutrient uptake and water absorption. These fungal networks connect individual plants, creating a vast underground communication system that allows trees to share resources and defend against pathogens. A healthy soil ecosystem is essential for maintaining woodland biodiversity; it provides the foundation for plant growth and supports a myriad of invertebrates that contribute to decomposition and nutrient cycling. Disturbing the soil through intensive management practices or pollution can disrupt these delicate relationships, impacting the entire ecosystem.

Let's not forget the often unseen world of insects. These tiny creatures are vital pollinators, herbivores, and decomposers, playing a crucial role in maintaining ecosystem function. A diverse insect community ensures efficient pollination of woodland plants, supporting fruit and seed production. Herbivorous insects, while sometimes viewed as pests, provide food for birds and other predators, contributing to the food web. Decomposers, such as beetles and springtails, break down leaf litter and dead wood, releasing nutrients back into the soil.

The presence of a variety of wildlife is also key. Birds contribute to seed dispersal and insect control, while mammals play important roles in herbivory, predation, and nutrient cycling. Different animal species have different habitat requirements and feeding preferences, contributing to the overall biodiversity of the woodland. A woodland with a healthy population of deer, for example, can support a diverse community of predators, such as foxes, owls, and birds of prey.

However, this delicate balance is easily disrupted. Habitat loss, fragmentation, pollution, and climate change all pose significant threats to woodland biodiversity. When habitats are destroyed or fragmented, species are forced to compete for limited resources, leading to declines in population sizes and local extinctions. Pollution can directly harm plants and animals, disrupting food webs and altering ecosystem function. Climate change is already impacting woodland ecosystems, altering species distributions, increasing the frequency of extreme weather events, and creating conditions that favor invasive species.

Protecting and enhancing woodland biodiversity requires a multifaceted approach. Conservation efforts must focus on preserving existing woodland habitats, restoring degraded areas, and creating corridors to connect fragmented landscapes. Sustainable management practices, such as selective logging and controlled burning, can help maintain woodland structure and promote biodiversity. Reducing pollution and mitigating climate change are also crucial for protecting woodland ecosystems in the long term.

Furthermore, fostering awareness and appreciation of woodland biodiversity is essential for inspiring conservation action. By educating the public about the importance of woodlands and the threats they face, we can encourage individuals to take steps to protect these valuable ecosystems. Supporting local conservation organizations, participating in citizen science projects, and making informed consumer choices are all ways that individuals can contribute to woodland conservation.

Understanding the intricate web of life within our woodlands is not merely an academic exercise; it is a fundamental step towards ensuring their survival. By recognizing the importance of biodiversity and taking action to protect it, we can safeguard these vital ecosystems for future generations. Remember, every species plays a role, and the loss of even a single thread can unravel the entire tapestry. So, next time you walk through a woodland, take a moment to appreciate the incredible diversity of life around you. Listen to the calls of the birds, observe the buzzing of the insects, and breathe in the scent of the earth. You are witnessing a masterpiece of nature, a testament to the power and resilience of life. Let's work together to ensure that future generations can experience the same wonder and awe.

2 Chapter 2: The Sun's Embrace

The sun, our woodland's primary energy source, paints a dynamic portrait across the forest, influencing every living thing within its realm. Without its life-giving rays, our vibrant plant relationships would cease to exist. Photosynthesis, the cornerstone of life, begins in the towering canopy, where leaves act as solar panels, capturing sunlight and transforming it into the sugars that fuel the tree's growth and survival.

The efficiency of this process is nothing short of remarkable. Pigments like chlorophyll within the leaves absorb specific wavelengths of light, primarily red and blue, while reflecting the green light that gives leaves their characteristic color. This captured energy drives a complex series of chemical reactions, combining carbon dioxide from the air with water absorbed from the soil to create glucose, a simple sugar. As a byproduct, oxygen is released back into the atmosphere, contributing to the air we breathe. Different tree species have evolved unique adaptations to maximize their photosynthetic efficiency. Some have broader leaves to capture more sunlight, while others have developed specialized pigments that can absorb a wider range of light wavelengths. In areas with dense canopies, leaves may even be arranged in a mosaic pattern, minimizing self-shading and ensuring that each leaf receives its fair share of sunlight. The entire process is a delicate balance, a symphony of light, water, and carbon dioxide orchestrated by the intricate machinery of the leaf. The glucose produced through photosynthesis serves as the foundation for the entire woodland food web, providing energy for the trees themselves, as well as the myriad of organisms that depend on them for sustenance. The process is not static; it changes throughout the year, responding to variations in light intensity, temperature, and water availability.

Light and shade dynamics in a woodland are far from uniform. The canopy acts as a filter, intercepting much of the incoming sunlight before it reaches the forest floor. This creates a mosaic of light and shade, where sun-drenched patches alternate with dimly lit areas. The intensity and duration of sunlight vary depending on the density of the canopy, the time of day, and the season. In spring, before the trees have fully leafed out, more sunlight penetrates to the forest floor, triggering a burst of growth in early-flowering plants. As the canopy fills in during summer, the understory becomes increasingly shaded, favoring plants adapted to low-light conditions. These variations in light availability profoundly influence plant distribution and growth patterns. Shade-tolerant species, such as ferns, mosses, and certain shrubs, thrive in the dimly lit understory, while sun-loving species are typically found in more open areas, like gaps created by fallen trees. The angle of the sun also plays a role. During the winter months, when the sun is lower in the sky, sunlight penetrates deeper into the woodland, providing essential light for evergreen plants and helping to melt snow cover. Understanding these light and shade dynamics is crucial for comprehending the complex interactions that shape plant communities within the woodland ecosystem. These dynamic shifts provide different plants to flourish depending on the time of the year. The balance of light and shade is not merely a physical phenomenon; it is a key driver of ecological processes, influencing everything from plant growth and competition to nutrient cycling and decomposition.

Beneath the canopy lies the forest floor, a world of perpetual twilight, where only a fraction of the sunlight filters through. This environment demands unique adaptations. Plants here often have larger leaves to maximize light capture, even if it is sparse. Their photosynthetic rates may be lower, but they are incredibly efficient at utilizing the available light. Many forest floor plants are ephemeral, meaning they emerge and flower early in the spring, before the canopy fully develops, taking advantage of the brief period of high light availability. Others are shade-tolerant, possessing specialized pigments that allow them to thrive in the dim light. The forest floor is not just a dark and shadowy place; it is a dynamic ecosystem teeming with life. Decomposers, such as fungi and bacteria, break down dead leaves and other organic matter, releasing nutrients back into the soil. These nutrients are then absorbed by plant roots, completing the cycle. The forest floor also provides habitat for a variety of animals, including insects, amphibians, and small mammals. Many of these creatures are adapted to the low-light conditions, possessing keen senses of smell and hearing to navigate their surroundings. The forest floor is a critical component of the woodland ecosystem, playing a vital role in nutrient cycling, decomposition, and habitat provision. The health of the forest floor is directly linked to the health of the entire woodland. Activities that disturb the forest floor, such as logging or excessive trampling, can disrupt the delicate balance of this ecosystem, leading to soil erosion, nutrient loss, and a decline in biodiversity. In essence, the forest floor is the foundation upon which the entire woodland community is built. My experience with plants, as well as understanding what woodland means, allowed me to learn this.

The impact of sunlight doesn't stop at photosynthesis and plant distribution. It also influences the temperature and humidity of the woodland. The canopy helps to moderate temperature fluctuations, providing shade during hot summer months and trapping heat during cold winter months. This creates a more stable microclimate beneath the canopy, which is beneficial for many woodland organisms. Sunlight also affects humidity levels. The canopy intercepts rainfall, reducing the amount of water that reaches the forest floor. This can create drier conditions in the understory, favoring plants that are adapted to drought. Sunlight also drives evapotranspiration, the process by which water evaporates from leaves and other plant surfaces. Evapotranspiration helps to cool the woodland and increase humidity levels. The interplay between sunlight, temperature, and humidity creates a complex microclimate that influences the distribution and abundance of plants and animals within the woodland. And I can't stress this enough. This intricate web of relationships underscores the importance of understanding the role of sunlight in shaping the woodland ecosystem.

2.1 Photosynthesis in the Canopy

The woodland canopy, a verdant ceiling woven from leaves and branches, orchestrates a symphony of sunlight. It's more than just the top layer of trees; it's the primary engine driving the entire woodland ecosystem through the remarkable process of photosynthesis. Understanding this process within the canopy is fundamental to grasping how the woodland functions.

Photosynthesis, at its core, is the conversion of light energy into chemical energy, fueling the growth and sustenance of plants. In the canopy, leaves act as solar panels, capturing sunlight with their chlorophyll pigments. The efficiency with which this capture occurs is influenced by a multitude of factors, including leaf angle, leaf area index (LAI), and the availability of essential resources such as water and nutrients. Trees in the upper canopy layers, directly exposed to sunlight, often have smaller, thicker leaves to minimize water loss, while those in the lower layers have larger, thinner leaves to maximize light absorption in the dim conditions. The intricate architecture of the canopy, with its complex arrangement of leaves, ensures that sunlight is intercepted and utilized effectively.

The journey of sunlight through the canopy is far from uniform. As sunlight penetrates the dense foliage, it encounters layers of leaves, each absorbing a portion of the light energy. This process of interception and absorption leads to a phenomenon known as light attenuation, where the intensity and spectral composition of light changes as it travels downwards. The uppermost layers receive the full spectrum of sunlight, while the lower layers are bathed in filtered, diffuse light enriched in green wavelengths due to the absorption of red and blue light by chlorophyll.

This differential light availability has profound implications for the plants growing beneath the canopy. Understory vegetation, including shrubs, herbs, and seedlings, must adapt to survive in the shade cast by the canopy trees. They often exhibit physiological adaptations, such as lower light compensation points and higher chlorophyll content, enabling them to photosynthesize efficiently even in low light conditions. The mosaic of sunflecks, fleeting patches of direct sunlight that penetrate the canopy, provides crucial opportunities for these shade-tolerant plants to boost their photosynthetic rates. The constant competition for light in this dense environment is a key driver of plant community structure and species distribution.

Moreover, the canopy plays a crucial role in regulating the woodland's microclimate. Through transpiration, the process of water movement through a plant and its evaporation from aerial parts such as leaves, trees release water vapor into the atmosphere, cooling the surrounding air and increasing humidity. This process also helps to draw water and nutrients from the soil up to the leaves. The canopy acts as a buffer, moderating temperature fluctuations and reducing wind speed, creating a more stable and sheltered environment for plants and animals. The extent of this buffering effect depends on the density and structure of the canopy, with denser canopies providing greater protection from extreme weather conditions.

The availability of water and nutrients also critically impacts photosynthetic rates within the canopy. Water stress can lead to stomatal closure, reducing the uptake of carbon dioxide, a vital ingredient for photosynthesis. Nutrient deficiencies, particularly nitrogen, can limit the production of chlorophyll, impairing light absorption. Trees in the canopy must effectively allocate resources to their leaves to optimize photosynthetic performance. Factors like soil quality, rainfall patterns, and the presence of mycorrhizal fungi influence the availability of these essential resources.

The efficiency of photosynthesis in the canopy is not static; it varies throughout the day and across seasons. During the day, photosynthetic rates typically peak in the morning and decline in the afternoon as temperatures rise and water stress increases. Seasonally, photosynthetic activity is highest during the growing season when temperatures are favorable and sunlight is abundant. In deciduous woodlands, the shedding of leaves in autumn marks the end of the photosynthetic period, with trees entering a dormant state to conserve energy. Understanding these temporal dynamics is essential for assessing the overall productivity of the woodland ecosystem.

The health of the canopy is an indicator of the overall health of the woodland. Factors such as air pollution, insect infestations, and fungal diseases can impair photosynthetic rates, leading to reduced growth and increased mortality. Deforestation and forest fragmentation can disrupt the canopy structure, altering light and shade dynamics and affecting the microclimate. Monitoring the health of the canopy, through techniques such as remote sensing and ground-based surveys, is crucial for detecting and addressing threats to woodland ecosystems.

Considering all aspects, the canopy is the photosynthetic powerhouse of the woodland, driving energy flow and shaping the environment for all other organisms. Its intricate architecture, physiological adaptations, and role in regulating microclimate make it a critical component of the ecosystem. By understanding the complexities of photosynthesis in the canopy, we can better appreciate the importance of preserving and managing these valuable woodland ecosystems for future generations. Furthermore, exploring the intricate relationships within the canopy and its connection to the entire ecosystem highlights the importance of studying these natural systems, driving further research to advance our knowledge and conservation efforts.

2.2 Light and Shade Dynamics

The interplay of light and shadow within a woodland is far from a static phenomenon; it's a vibrant, ever-shifting dance that dictates much of what occurs on the forest floor and up into the canopy. The sun, the prime driver of life in nearly all ecosystems, filters through the leaves, creating a mosaic of intense brightness and deep, cool shade. This isn't just about illumination; it's about energy distribution, temperature regulation, and the very survival of countless plant species. I always find the variability to be one of the most intriguing things.

Consider the sun's journey from dawn to dusk. At sunrise and sunset, the light comes in at a low angle, casting long, dramatic shadows that stretch across the forest floor. During these times, even normally shaded areas receive a fleeting bath of direct sunlight, triggering brief bursts of photosynthetic activity in plants that are typically starved of it. These moments can be crucial for the growth of certain understory species or the germination of seeds that require a light trigger. As the sun climbs higher in the sky, the intensity of the light increases, but the canopy begins to exert its influence. Leaves act as natural filters, absorbing some wavelengths of light and reflecting or transmitting others. This process alters the spectral composition of the light that reaches the lower layers of the woodland, favoring shade-tolerant plants that have adapted to thrive under these conditions.

The density of the canopy is a major factor. In a dense, closed-canopy woodland, only a small percentage of the sunlight reaches the forest floor – often as little as 1-2%. This dim light environment favors plants with specialized adaptations, such as large leaves to maximize light capture, or the ability to utilize far-red light, which penetrates the canopy more effectively than other wavelengths. Think of the delicate ferns and mosses that carpet the forest floor, or the intricate patterns of leaf veins designed to channel every precious photon towards the plant's photosynthetic machinery. Now contrast that with an open-canopy woodland, where gaps in the tree cover allow significantly more sunlight to penetrate. These sun-drenched patches become havens for plants that demand high light levels, like wildflowers and grasses, creating localized areas of intense biodiversity. The variation in light availability creates a patchwork of microhabitats.

The time of year drastically influences the light climate. During the growing season, the canopy is at its fullest, casting the deepest shadows. This is when the understory plants race to complete their life cycles before the canopy closes in completely. Many spring ephemerals, for instance, bloom and set seed before the trees leaf out, taking advantage of the brief period of high light availability. As autumn arrives, the leaves begin to fall, gradually opening up the canopy and allowing more light to reach the forest floor. This autumnal flush of sunlight can stimulate late-season growth in some plants and prepare them for winter dormancy. Then winter arrives with a bare canopy.

The impact of light and shade isn't limited to plants. Animals, too, are affected by the light dynamics of the woodland. Many insects, for example, are highly sensitive to light levels and use the shaded areas for refuge from predators or to regulate their body temperature. Birds often choose their nesting sites based on light availability, seeking out areas where they can effectively camouflage their nests or where their young can receive adequate sunlight for development. Mammals, especially nocturnal species, rely on the shadows for cover and protection. Understanding these light-related patterns is crucial for appreciating the complex web of interactions that define a woodland ecosystem.

Consider a specific example: a patch of wild garlic growing on the forest floor. In early spring, before the trees have leafed out, the garlic receives ample sunlight, allowing it to rapidly grow and produce its characteristic pungent leaves. As the canopy closes in, the garlic's growth slows down, but it has already stored enough energy to survive the shaded summer months. This seasonal adaptation is a direct result of the light dynamics within the woodland. The angle of sunlight can also affect the temperature of the soil. Areas exposed to direct sunlight will warm up faster than shaded areas, influencing seed germination and root growth. The moisture content of the soil is also affected, with sunnier areas drying out more quickly than shaded areas. These microclimatic variations further contribute to the diversity of plant and animal life within the woodland. So many factors come into play. I often observe how changes in land use can dramatically alter the light environment of a woodland. Deforestation, for example, removes the canopy altogether, exposing the forest floor to intense sunlight and drastically altering the temperature and moisture regimes. This can lead to the loss of shade-tolerant species and the invasion of sun-loving plants, transforming the character of the woodland. Selective logging, on the other hand, can create gaps in the canopy, increasing light availability in localized areas and promoting the growth of certain understory plants. But it's important to manage these interventions carefully to avoid disrupting the delicate balance of the ecosystem.

For anyone interested in studying woodland ecosystems, understanding light and shade dynamics is absolutely fundamental. It's a complex and fascinating subject that requires careful observation and a deep appreciation for the interplay of sunlight, plants, and animals. So next time you find yourself in a woodland, take a moment to notice the patterns of light and shadow. Observe how the sunlight filters through the leaves, creating a mosaic of brightness and shade. Look at the plants growing in different light conditions and consider how they have adapted to their environment. Listen to the sounds of the animals that rely on the shadows for protection. By paying attention to these details, you can gain a deeper understanding of the intricate and beautiful world of the woodland. It is one of my favourite things.

2.3 The Forest Floor

Beneath the towering canopy, a world teeming with life unfolds, often unseen and unappreciated: the forest floor. This is where the dance of decomposition plays out, where seeds patiently await their moment, and where a myriad of creatures find sustenance and shelter. It is, in essence, the foundation upon which the entire woodland ecosystem rests.

The first thing that strikes you is the sheer complexity of the layer. It's not just dirt and fallen leaves; it's a mosaic of organic matter in various stages of decay. A vibrant carpet of fallen leaves, twigs, and decaying wood covers the mineral soil beneath. This layer, known as the litter layer, is the starting point for a fascinating process. As these materials decompose, they release vital nutrients back into the soil, nourishing the plants above. Fungi, bacteria, and a host of invertebrates work tirelessly, breaking down complex organic molecules into simpler forms that plants can readily absorb. This constant cycle of decay and renewal is the lifeblood of the woodland, driving nutrient cycling and maintaining the overall health of the ecosystem. What lies beneath this layer is equally crucial, with each element contributing to the intricate balance of the woodland.

One of the most remarkable aspects of the forest floor is its role as a seed bank. Countless seeds from trees, shrubs, and herbaceous plants lie dormant in the soil, waiting for the right conditions to germinate. Some seeds require a period of cold stratification, where they must experience a period of cold temperatures before they can sprout. Others need a burst of light or a disturbance in the soil to trigger germination. This seed bank is a reservoir of potential, ensuring that the woodland can regenerate even after disturbances such as fires or windstorms. The composition of the seed bank can vary greatly depending on the location, the species present in the surrounding area, and the history of the site. Areas with frequent disturbances may have a higher proportion of fast-growing, opportunistic species, while more stable areas may be dominated by slower-growing, shade-tolerant species. The forest floor is also a critical habitat for a diverse array of animals. Insects, mites, spiders, and other invertebrates thrive in the moist, dark environment, feeding on decaying organic matter and preying on each other. These creatures play a vital role in the decomposition process and contribute to soil aeration and nutrient cycling. Larger animals, such as amphibians, reptiles, and small mammals, also rely on the forest floor for shelter, foraging, and nesting. Salamanders, for example, are particularly sensitive to changes in the forest floor environment, as they require moist conditions to survive. Earthworms, often overlooked, are ecosystem engineers, tirelessly aerating the soil and breaking down organic matter. They create tunnels that improve drainage and allow plant roots to penetrate deeper into the soil.

Moreover, the amount of light that filters through the canopy has a profound impact on the plant life that can survive on the forest floor. In areas with dense canopies, only a fraction of the sunlight reaches the ground, favoring shade-tolerant species such as mosses, ferns, and certain wildflowers. These plants have adapted to low-light conditions by developing larger leaves, specialized pigments to capture what light is available, and efficient photosynthetic pathways. In contrast, areas with more open canopies receive more sunlight, allowing for a greater diversity of plant life, including sun-loving wildflowers and grasses. These differences in light availability create distinct microhabitats within the forest floor, supporting a wide range of species. It's a continuous interplay between the sun, the trees, and the life below.

Now, consider the impact of human activities on the forest floor. Soil compaction from logging or recreational use can damage the delicate soil structure, reducing aeration and water infiltration. This can hinder plant growth and make it difficult for seedlings to establish. The introduction of invasive species can also disrupt the natural balance of the forest floor, outcompeting native plants and altering the decomposition process. Pollution, such as acid rain or heavy metal contamination, can harm soil organisms and affect nutrient cycling. Understanding these threats is crucial for developing effective conservation strategies.

When you think about the health of a woodland, look down, too. Protecting the forest floor is essential for maintaining the overall health and resilience of the entire ecosystem. Sustainable forestry practices, such as selective harvesting and minimizing soil disturbance, can help to maintain the integrity of the forest floor. Restoring degraded areas by planting native vegetation and removing invasive species can also improve soil health and habitat quality. Furthermore, educating the public about the importance of the forest floor and encouraging responsible recreational use can help to minimize human impacts.

One practical thing anyone can do is to be mindful of their actions when visiting a woodland. Stay on designated trails to avoid compacting the soil and disturbing sensitive vegetation. Avoid collecting plants or other natural materials, as this can disrupt the ecosystem. Report any signs of invasive species to local authorities. And most importantly, take the time to appreciate the beauty and complexity of the forest floor – the hidden world beneath our feet.

Finally, reflecting on the delicate interplay of life and decay, sunlight and shadow, and the myriad of organisms that call it home, the forest floor is not just a layer of dirt and leaves; it's a dynamic and essential component of the woodland ecosystem. It's a place where the past meets the future, where nutrients are recycled, and where life finds a way, even in the darkest corners. By understanding and appreciating the forest floor, we can better protect this valuable resource and ensure the health and resilience of our woodlands for generations to come. I believe it's our duty to be stewards of this hidden realm, ensuring its continued vitality and its contribution to the overall health of our planet.

3 Chapter 3: Plant Communities

Stepping into a woodland is like entering a multi-layered world, each stratum teeming with a unique cast of plant life. Understanding how these plants organize themselves into communities is fundamental to grasping the overall health and function of the ecosystem. The interplay of light, water, nutrients, and even animal activity shapes these communities, creating a dynamic tapestry of life.

Dominant tree species define the character of a woodland. In many northern regions, you might find towering oaks, their broad canopies casting dappled shade below. Beech trees, with their smooth, grey bark, are other common players, sometimes forming extensive, single-species stands. Conifers like pines and firs dominate in cooler, more acidic environments. Each of these trees exerts a profound influence on the environment beneath them. Oak leaves, for example, decompose slowly, creating a rich leaf litter that supports a particular set of fungi and invertebrates. Beech trees, on the other hand, cast a dense shade that limits the growth of many understory plants.

Consider the American Chestnut, once a dominant species in eastern North American woodlands, decimated by a fungal blight. Its absence drastically altered the composition of those forests, highlighting the ripple effects of losing a keystone species.

Below the canopy lies the understory, a realm of shrubs, saplings, and smaller trees vying for sunlight. Dogwoods, with their beautiful spring blooms, are common in many eastern woodlands. Hazel, with its edible nuts, provides sustenance for wildlife. These plants are adapted to the lower light levels and often possess broader leaves to capture as much sunlight as possible. They create a vertical structure that provides habitat for birds, insects, and small mammals. Think about the thorny thickets created by some shrubs, offering protection to nesting birds from predators. These understory plants also play a crucial role in nutrient cycling, their roots helping to stabilize the soil and prevent erosion.

At ground level, a carpet of life unfolds. Ground cover plants, mosses, and lichens form a mosaic of textures and colors. Wildflowers like trilliums and violets add splashes of color in the spring. Mosses cling to rocks and tree trunks, creating miniature ecosystems. Lichens, a symbiotic partnership between fungi and algae, can be found on almost any surface, from bare soil to the branches of ancient trees. These ground cover plants help to retain moisture, prevent soil erosion, and provide habitat for invertebrates. Picture the intricate patterns formed by mosses on a fallen log, a tiny world teeming with microscopic life.

Ephemeral blooms are the woodland's fleeting jewels. These wildflowers, often blooming in early spring before the trees leaf out, take advantage of the increased sunlight reaching the forest floor. Spring beauties, trout lilies, and Dutchman's breeches are among the many species that grace the woodland with their delicate flowers. They attract pollinators like bees and butterflies, playing a crucial role in the reproduction of these plants. These ephemeral blooms are a reminder of the dynamic nature of the woodland, a burst of life before the canopy closes in.

Invasive species pose a significant threat to the integrity of woodland plant communities. These non-native plants, often introduced accidentally or intentionally, can outcompete native species for resources like sunlight, water, and nutrients. Garlic mustard, for example, can quickly spread and dominate the forest floor, inhibiting the growth of native wildflowers and tree seedlings. Japanese honeysuckle, with its aggressive vines, can smother trees and shrubs. These invasive species can disrupt the delicate balance of the ecosystem, reducing biodiversity and altering habitat structure. Managing invasive species requires ongoing effort, including manual removal, herbicide application, and biological control. Understanding the pathways of introduction and spread is crucial for preventing future invasions. Imagine the impact of a single invasive vine, gradually engulfing a native tree, ultimately leading to its demise and the loss of habitat for the animals that depend on it.

The story of these plant communities goes further than just their presence and function. The interaction between all of them results in more than just a collection of plants, it is a whole, an ecosystem that must be preserved.

3.1 Dominant Tree Species

To truly grasp the essence of a woodland, we must first acknowledge the titans that define it: the dominant tree species. These are the trees that, through sheer size, longevity, and competitive prowess, dictate the structure, resources, and overall character of the woodland environment. They are the foundation upon which all other life in the ecosystem is built. Identifying these species is not merely about knowing their names; it's about understanding their individual roles and how they contribute to the larger woodland community.

Consider the mighty oak, a common dominant tree in many temperate woodlands. Oaks, belonging to the Quercus genus, are not just impressive in stature; they are ecological powerhouses. They provide copious amounts of leaf litter, which decomposes to enrich the soil, supporting a diverse community of invertebrates and fungi. Their acorns are a vital food source for numerous animals, from squirrels and deer to birds like jays and woodpeckers. The oak's dense canopy creates shade, influencing the understory vegetation and microclimate below. Even in death, a fallen oak continues to contribute, providing habitat for insects, fungi, and eventually returning its nutrients to the soil. Different oak species will favor different habitats, with white oaks preferring well-drained soils and red oaks tolerating slightly wetter conditions. This subtle variation can influence the overall composition of the woodland.

Then there's the beech, another common dominant species, particularly in European woodlands. Beeches, of the Fagus genus, are known for their smooth, gray bark and dense canopies. Unlike oaks, beech leaf litter decomposes slowly, leading to a different soil composition and impacting which plants can thrive beneath them. The dense shade cast by beech trees can limit the growth of other species, making them strong competitors. Beech nuts, like acorns, are an important food source for wildlife. Furthermore, beech trees are particularly susceptible to certain fungal diseases, which can create gaps in the canopy and alter the light regime in the understory, creating opportunities for other plants to establish. Understanding the impact of these diseases on dominant species is crucial for understanding the long-term dynamics of the woodland.

Beyond oaks and beeches, other tree species can dominate specific woodland types. Pines, of the Pinus genus, are often dominant in coniferous woodlands, particularly in areas with poorer soils or harsher climates. Their needle-like leaves are adapted to conserve water, allowing them to thrive in drier conditions. Pine forests support a unique community of plants and animals adapted to their specific environment, including specialized fungi that form mycorrhizal associations with pine roots. Alders, belonging to the Alnus genus, are often found in wetter areas, such as along streams and rivers. They have the unique ability to fix nitrogen in the soil, enriching it and supporting a diverse community of plants. Alders also provide important habitat for insects and other invertebrates, contributing to the food web of the woodland ecosystem.

Identifying dominant tree species requires careful observation and a basic understanding of tree identification. Start by looking at the overall shape and size of the trees. Are they tall and slender, or broad and spreading? Examine the bark carefully, noting its color, texture, and pattern. Look at the leaves: their shape, size, arrangement on the branches, and whether they are simple or compound. Collect fallen leaves and twigs to examine them more closely. Using a field guide or online resources, compare your observations to descriptions and illustrations of different tree species. Consider the habitat: is the woodland dry or wet, sunny or shady? Certain tree species are more likely to be found in specific habitats.

Beyond simple identification, understanding the role of each dominant species involves studying their ecological characteristics. Research their growth rates, lifespan, reproductive strategies, and susceptibility to diseases and pests. Observe how they interact with other plants and animals in the woodland ecosystem. Are they providing food or shelter for wildlife? Are they competing with other trees for resources? Are they influencing the soil composition or water availability? Consider the history of the woodland. Has it been subject to logging, grazing, or other disturbances? These factors can influence the composition and structure of the dominant tree species.

The presence or absence of particular dominant species can tell us a great deal about the health and history of a woodland. For example, a decline in oak populations may indicate habitat loss, climate change, or the presence of invasive species. An increase in the abundance of a particular tree species may suggest a change in environmental conditions or a shift in management practices. By monitoring the populations of dominant tree species, we can gain valuable insights into the overall health and resilience of the woodland ecosystem.

Ultimately, understanding dominant tree species is fundamental to understanding the whole woodland. They are the architects of the environment, shaping the conditions that support all other life. By learning to identify and appreciate these magnificent trees, we can gain a deeper understanding of the complex and interconnected world of the woodland ecosystem. And this, in turn, allows us to better protect and manage these valuable resources for future generations. They are not just trees; they are the keystone species that hold the entire ecosystem together. Observe them closely, and the secrets of the woodland will slowly reveal themselves.

3.2 Understory Vegetation

The understory is a captivating layer in any woodland. I think of it as a dynamic space, squeezed between the towering canopy and the humble forest floor, a place where light and shadow dance and where a unique cast of plants thrives. It is far from just a collection of random shrubs and saplings. It's a distinct community, shaped by the specific light conditions and the interactions with other species.

Think of the dominant trees towering above. Their canopy dictates much of what happens beneath. They filter sunlight, casting dappled shade that only certain plants can tolerate. This is where shade-tolerant shrubs like dogwood, hazel, and witch hazel come into their own. You will also find younger trees, saplings patiently waiting for an opportunity to grow larger if a gap in the canopy opens up. Each plant has its own strategy for survival in this environment.

The plants that call this understory home are not just miniature versions of their taller, canopy-dwelling relatives. They exhibit unique adaptations to the reduced light levels. Their leaves, for example, tend to be larger and thinner than those of canopy trees, maximizing light absorption. Some, like spring ephemerals, are only active for a short period in early spring, taking advantage of the brief window of full sunlight before the canopy leaves fully emerge. Others might use the few direct sunlight opportunities to blossom.

Beyond their physical adaptations, these plants contribute significantly to the woodland ecosystem. They provide food and shelter for a variety of animals, from insects and birds to small mammals. Many species of birds build their nests in the understory shrubs, finding protection from predators and the elements. Herbivores, in turn, browse on the leaves and twigs, helping to shape the structure and composition of the understory vegetation. The fallen leaves decompose, enriching the soil and providing nutrients for other plants. It's a cycle of life, intricately interwoven.

Furthermore, understanding the understory is crucial for forest management and conservation. This layer can act as a barrier against invasive species, with healthy, diverse understory communities capable of outcompeting newcomers. It also plays a critical role in forest regeneration. A well-developed understory, with its mix of seedlings and saplings, ensures that the woodland can recover quickly after disturbances, such as fire or logging.

However, the understory is not immune to threats. Overgrazing by deer can severely damage or eliminate certain plant species, leading to a decline in biodiversity. Invasive plants, if they manage to establish themselves, can quickly outcompete native species, transforming the structure and function of the understory. Climate change, with its altered precipitation patterns and increased frequency of extreme weather events, can also have significant impacts on the delicate balance of this layer.

To protect and conserve the understory, we need to implement responsible forest management practices. This includes controlling deer populations, preventing the introduction and spread of invasive species, and mitigating the effects of climate change. We can also promote the growth and diversity of the understory through careful thinning of the canopy, allowing more light to reach the forest floor.

Consider participating in citizen science projects that monitor the health and composition of woodland ecosystems. By collecting data on plant species, abundance, and distribution, we can gain a better understanding of the changes occurring in the understory and inform effective conservation strategies. Educating others about the importance of this often-overlooked layer is also vital. Share your knowledge with friends, family, and community members, fostering a greater appreciation for the complexity and beauty of woodland ecosystems.

As I think of it, the understory is not just a static component of the woodland. It's a dynamic and ever-changing community, constantly responding to the interplay of light, shade, competition, and disturbance. Its health and diversity are essential for the overall health and resilience of the entire ecosystem. By understanding and appreciating this vital layer, we can work together to protect it for future generations.

So, the next time you wander through a woodland, take a moment to look beyond the towering trees and notice the understory. Observe the different plant species, their unique adaptations, and the intricate relationships they form with other organisms. You might be surprised by what you discover. The understory is a world unto itself, full of hidden wonders and ecological significance. It is a vital piece of the puzzle, contributing to the richness and complexity of the woodland ecosystem. Protecting it ensures the health and vitality of the entire forest.

3.3 Ground Cover Plants

Ground cover plants, often overlooked, are the unsung heroes of the woodland floor. These resilient plants form a living carpet, playing vital roles in soil health, nutrient cycling, and overall ecosystem stability. They're not just pretty faces; they're functional components that contribute significantly to the health of the woodland. For those just starting to delve into the wonders of woodland ecosystems, understanding the importance and selection of ground cover plants is a crucial step.

Selecting the right ground cover requires careful consideration of several factors, the most important being light availability. As we discussed, light conditions on the forest floor are drastically different than in the canopy. Observe how much sunlight reaches different areas of your woodland. Deep shade is best suited for species like wild ginger (Asarum canadense) or partridgeberry (Mitchella repens), while areas with dappled sunlight can support a wider range of options such as Pennsylvania sedge (Carex pensylvanica) or creeping phlox (Phlox stolonifera). Soil type is another critical consideration. Is your soil acidic or alkaline? Does it drain well, or does it tend to stay wet? Matching the right ground cover to your soil conditions will significantly increase its chances of thriving. For example, ferns generally prefer moist, acidic soil, while sedums can tolerate drier, more alkaline conditions.

Beyond these environmental factors, consider the intended function of your ground cover. Are you looking to suppress weeds, prevent erosion, or simply add visual interest? Certain ground covers are more effective at weed suppression than others. For example, pachysandra (Pachysandra terminalis) forms a dense mat that can effectively smother weeds, but it can also be invasive in some areas. Native ground covers like wild strawberry (Fragaria virginiana) not only provide beautiful ground cover but also offer a food source for wildlife. Understanding your goals will help you narrow down your options.

Once you've assessed your site conditions and determined your objectives, it's time to explore the diverse world of ground cover plants. Native species are generally the best choice, as they are well-adapted to the local climate and soil conditions, and they provide valuable habitat for native wildlife. Wild ginger, mentioned earlier, is a beautiful and shade-tolerant option with heart-shaped leaves and intriguing, though often hidden, flowers. Partridgeberry is another excellent choice for shade, with its delicate evergreen foliage and bright red berries that persist through the winter. For areas with slightly more sunlight, consider Pennsylvania sedge, a grass-like plant that provides excellent erosion control and adds a soft, textured look to the landscape. Creeping phlox, with its vibrant spring blooms, is a great option for adding color and attracting pollinators.

When planting ground cover, proper preparation is essential. Remove any existing weeds or debris from the planting area. Amend the soil with compost or other organic matter to improve its fertility and drainage. Dig a hole that is slightly larger than the root ball of the plant, and gently loosen the roots before planting. Water thoroughly after planting, and continue to water regularly until the plants are established. Mulching with wood chips or shredded bark can help retain moisture, suppress weeds, and regulate soil temperature.

Maintaining ground cover plants typically involves minimal effort. Regular watering may be necessary during dry periods, especially in the first year after planting. Pruning may be required to control the spread of certain species, such as pachysandra or creeping phlox. Remove any dead or diseased foliage to prevent the spread of disease. Fertilizing is generally not necessary for native ground covers, but a light application of compost or organic fertilizer in the spring can help boost growth.

One common mistake that beginners make is planting ground cover too sparsely. Ground cover plants need to be planted close enough together to form a dense mat, which will effectively suppress weeds and prevent erosion. Check the plant label for recommended spacing, and don't be afraid to plant a little closer together if you want faster coverage. Another mistake is neglecting to water the plants regularly after planting. Ground cover plants need consistent moisture to establish themselves, especially during hot or dry weather.

Beyond the practical benefits, ground cover plants can also enhance the aesthetic appeal of your woodland. They can add color, texture, and visual interest to the landscape, creating a more inviting and enjoyable space.

Consider incorporating a variety of ground cover species to create a more diverse and dynamic landscape. Combine different textures and colors to create a visually appealing tapestry. Use ground cover plants to define pathways, create borders, or highlight specific features in your woodland. Finally, remember that the success of your ground cover planting depends on careful observation and ongoing adaptation. Pay attention to how your plants are growing, and adjust your care practices accordingly. If you notice signs of disease or pest problems, take action promptly to prevent further damage. Be patient, as it may take several years for ground cover plants to fully establish themselves and form a dense mat. With a little planning and effort, you can create a beautiful and functional ground cover layer that will enhance the health and beauty of your woodland for years to come. Don't be afraid to experiment with different species and techniques to find what works best for your particular site and preferences. The woodland floor is a dynamic and ever-changing environment, and your ground cover planting should reflect that. Embrace the natural processes and learn from your experiences, and you'll be well on your way to creating a thriving woodland ecosystem. Consider, too, that certain ground covers contribute to insect life – specific butterflies, for example, need host plants that may be ground covers.

3.4 Ephemeral Blooms

Emerging from the still-chilled earth after the long winter slumber, ephemeral blooms are a fleeting spectacle of color and life that marks the transition from dormancy to full spring awakening in woodlands. These short-lived wildflowers take advantage of a unique window of opportunity before the canopy fully leafs out, monopolizing the sunlight that streams through the bare branches to fuel their rapid growth, flowering, and seed production. Their existence is a race against time, a burst of activity squeezed into a few precious weeks before the dense shade of the established trees reclaims the forest floor.

These plants aren't merely pretty faces; they are vital components of the woodland ecosystem. They provide early-season nectar and pollen resources for emerging pollinators like bees and flies, sustaining these crucial insects as they awaken from their overwintering states. Their foliage offers a tender, nutritious food source for various woodland creatures, from deer browsing on the leaves to small mammals foraging for the bulbs and roots. The ephemeral bloom cycle is intricately interwoven with the life cycles of other organisms, creating a delicate web of interdependence.

Understanding the adaptations that allow these plants to thrive in their unique niche is key to appreciating their ecological role. One crucial adaptation is their ability to store energy in underground structures like bulbs, corms, or rhizomes. These structures act as nutrient reservoirs, accumulating resources during the brief period of sunlight and allowing the plants to rapidly emerge and flower in the spring. Without this energy storage, they would be unable to compete with the established trees and shrubs for resources.

Their flowering strategies are equally remarkable. Many ephemeral blooms exhibit vibrant colors and intricate patterns that attract specific pollinators. For example, the bright yellow petals of trout lilies and the nodding blossoms of bloodroot are signals to early-emerging bees, promising a valuable source of nectar and pollen. Some species, like spring beauties, rely on small, specialized bees that are particularly well-suited to pollinating their flowers. The timing of their flowering is also critical, synchronized with the emergence of their primary pollinators to ensure successful reproduction.

The life cycle of an ephemeral bloom is a carefully choreographed dance with the changing seasons. After flowering and pollination, they quickly set seed. These seeds often have elaiosomes, fleshy appendages that attract ants. The ants carry the seeds back to their nests, consuming the elaiosomes and inadvertently dispersing the seeds throughout the woodland. This ant dispersal, known as myrmecochory, is a crucial mechanism for expanding the range of these plants and colonizing new areas.

Following seed dispersal, the aboveground parts of the ephemeral blooms wither and die back, leaving behind only the underground storage structures. These structures remain dormant throughout the summer and fall, patiently accumulating resources for the next spring's burst of growth. This period of dormancy is just as important as the active growth phase, allowing the plants to conserve energy and survive the harsh conditions of the shaded forest floor.

However, the existence of these delicate plants is threatened by a variety of factors. Habitat destruction, due to logging, development, and agriculture, is a major concern. When woodlands are fragmented or destroyed, ephemeral blooms lose their habitat and their populations decline. Invasive species also pose a significant threat. Non-native plants can outcompete native ephemeral blooms for resources, disrupting the delicate balance of the ecosystem. Overgrazing by deer can also damage ephemeral bloom populations, as deer often browse on their tender foliage and flowers.

Climate change is another emerging threat. Changes in temperature and precipitation patterns can disrupt the timing of ephemeral bloom cycles, potentially leading to mismatches with their pollinators. Warmer temperatures may cause them to emerge earlier in the spring, before their pollinators are active, or may lead to increased competition from other plants. Shifts in precipitation patterns can also affect their growth and reproduction.

Protecting these plants requires a multi-faceted approach. Conservation efforts should focus on preserving and restoring woodland habitats. This includes protecting existing woodlands from development, implementing sustainable forestry practices, and controlling invasive species. It also involves managing deer populations to reduce overgrazing pressure. Educating the public about the importance of ephemeral blooms and the threats they face is also essential. By raising awareness, we can encourage responsible stewardship of these fragile ecosystems.

I remember stumbling upon a breathtaking display of Dutchman's breeches during a spring hike. The delicate, pantaloon-shaped flowers seemed to float above the forest floor, illuminated by the dappled sunlight. It was a moment of pure magic, a reminder of the beauty and resilience of the natural world. It strengthened my resolve to advocate for the protection of these incredible plant communities.

One can also support local conservation organizations that are working to protect woodland habitats. Participating in citizen science projects that monitor ephemeral bloom populations can contribute valuable data to research efforts. Even simple actions, such as avoiding trampling on wildflowers when hiking, can help to minimize our impact on these delicate ecosystems. We also can create suitable habitat in our own gardens by planting native ephemeral blooms.

Ephemeral blooms are more than just beautiful wildflowers; they are integral components of the woodland ecosystem. Their short-lived existence is a testament to the power of adaptation and the intricate connections that bind together the natural world. By understanding and appreciating their ecological role, and by taking action to protect their habitats, we can ensure that these ephemeral beauties continue to grace our woodlands for generations to come. Their delicate presence is a reminder of the importance of conservation and the need to protect the biodiversity of our planet.

3.5 Invasive Species

They arrive stealthily, silently disrupting the balance of the woodland, invaders with no invitation. These unwanted guests, the invasive species, can choke the life out of native plants and drastically alter the structure and function of our beloved woodlands. It's a slow burn, a creeping takeover that often goes unnoticed until the damage is extensive. Understanding their impact and how to manage them is a critical task for anyone interested in preserving the integrity of these valuable ecosystems.

The defining characteristic of an invasive species is its ability to aggressively spread outside its natural range. This is often due to a lack of natural predators or diseases that would keep it in check in its native habitat. Without these limitations, it can outcompete native species for resources like sunlight, water, and nutrients. The effects are far-reaching, impacting everything from the composition of plant communities to the availability of food and shelter for wildlife. Picture a dense carpet of garlic mustard smothering the forest floor, preventing native wildflowers from germinating and stealing vital resources from tree seedlings. Imagine a vine, like Japanese honeysuckle, twining around and strangling a mature oak, eventually causing its demise. These are just glimpses of the devastation invasive species can inflict.

Recognizing the enemy is the first step in combating their spread. Many invasive plants possess characteristics that make them easy to identify. Some, like multiflora rose, are readily recognizable by their thorny stems and clusters of white flowers. Others, like Japanese knotweed, are known for their rapid growth and bamboo-like stems. Still others, such as the aforementioned garlic mustard, will have the characteristic smell of garlic when their leaves are crushed. Familiarizing yourself with the common invaders in your region is essential. Consult with local conservation organizations, botanical gardens, or forestry agencies to obtain field guides and resources that can aid in identification. Online databases and apps can also be valuable tools, providing photos, descriptions, and distribution maps.

Once you can identify the invaders, the next step is understanding how they spread. Many invasive plants are prolific seed producers, with seeds that can be dispersed by wind, water, or animals. Some species can also reproduce vegetatively, through rhizomes or root fragments. This means that even a small piece of the plant can potentially give rise to a new infestation. Human activities also play a significant role in the spread of invasive species. Seeds can be transported on clothing, shoes, or equipment. Construction, logging, and other land-disturbing activities can create ideal conditions for invasive plants to establish themselves. Even seemingly innocuous activities, such as planting non-native ornamentals in gardens, can contribute to the problem.

Control methods vary depending on the species, the size of the infestation, and the ecological context. For small infestations, manual removal may be the most effective approach. This involves hand-pulling the plants, taking care to remove as much of the root system as possible. It's important to wear gloves and appropriate clothing to protect yourself from thorns, sap, or other irritants. Persistence is key, as many invasive plants will re-sprout from remaining roots or seeds. Regular monitoring and follow-up treatments are often necessary to achieve long-term control.

When manual removal is not feasible, other methods may be necessary. Herbicides can be effective for controlling large infestations, but they should be used with caution, as they can also harm non-target plants. It's important to select a herbicide that is specifically designed for the target species and to follow all label instructions carefully. In some cases, biological control agents, such as insects or pathogens that attack the invasive plant, may be available. However, biological control should only be used after careful research and evaluation, to ensure that the agent does not pose a threat to native species.

Prevention is always the best strategy when it comes to invasive species. Preventing their introduction and spread in the first place is far more effective and cost-effective than trying to control them once they have become established. Educating yourself and others about the risks posed by invasive species is crucial. Avoid planting non-native species in your garden or yard. When hiking or working in the woods, clean your boots, clothing, and equipment before leaving the area to prevent the spread of seeds. Support local conservation organizations and participate in volunteer efforts to remove invasive plants from woodlands and other natural areas.

Restoring woodland ecosystems that have been degraded by invasive species can be a challenging but rewarding process. Simply removing the invasive plants is often not enough to restore the ecosystem to its former state. The soil may be depleted of nutrients, the native plant community may be fragmented, and the wildlife may have been displaced. Restoration efforts may involve reintroducing native plants, improving soil conditions, and creating habitat for wildlife. It's a long-term commitment that requires careful planning, ongoing monitoring, and adaptive management.

Consider the bigger picture and how seemingly small actions have ripple effects in complex ecosystems. The purchase of one non-native plant can become a disaster for the surrounding woods. Our collective choices determine whether we safeguard these beautiful, intricate environments or inadvertently contribute to their decline.

4 Chapter 4: Symbiotic Relationships

Imagine a web woven beneath our feet, a silent, subterranean exchange humming with life. This isn't the realm of earthworms and grubs alone; it's a vast network of fungal filaments, the mycorrhizae, embracing the roots of almost every plant in the woodland. These aren't parasites, but partners in a crucial exchange, a symbiotic dance that sustains the entire ecosystem. Think of it as the plant's extended root system, a microscopic exploration team reaching far beyond the plant's own capabilities. The fungi, in return for sugars produced by the plant through photosynthesis, provide vital nutrients like phosphorus and nitrogen, unlocking them from the soil in forms the plant can readily absorb. I've spent countless hours carefully excavating soil samples, meticulously teasing apart the delicate fungal threads clinging to root tips. Each sample reveals a slightly different story, a unique partnership tailored to the specific needs of the plant and the available resources in the soil. This is why clear-cutting a forest can be so devastating; it disrupts this intricate network, leaving the remaining plants vulnerable.

Consider the delicate orchids, often found nestled in the dappled shade of the woodland floor. Many orchid species are almost entirely dependent on mycorrhizal fungi for germination. Their seeds are tiny, lacking the stored energy necessary to sprout on their own. Instead, they rely on a fungal partner to provide the initial nourishment, kickstarting their growth. The fungus penetrates the seed, delivering sugars and nutrients until the young orchid can produce its own food through photosynthesis. This relationship can persist throughout the orchid's life, with the fungus continuing to provide essential support. Furthermore, trees benefit in remarkable ways; some research even suggests that mycorrhizal networks allow trees to communicate with each other, warning of threats like insect infestations or drought. A tree under attack might send chemical signals through the fungal network, prompting its neighbors to activate their defenses. Imagine the entire woodland connected by this vast, invisible internet, sharing information and resources in a constant flow. This concept, while still under investigation, highlights the profound interconnectedness of the woodland ecosystem.

Now, let's turn our attention to the vibrant world above ground, where flowers beckon with color and fragrance. Pollination, the transfer of pollen from the male part of a flower (the stamen) to the female part (the pistil), is essential for plant reproduction, and the woodland is a stage for a myriad of ingenious pollination strategies. Wind pollination is common among many tree species, relying on the random dispersal of pollen grains by air currents. These trees often produce copious amounts of pollen, creating the familiar yellow dust that coats everything in spring. However, many woodland plants rely on animals, primarily insects, to carry pollen from flower to flower. These plants entice their pollinators with a variety of rewards, including nectar, pollen, and even shelter. The bright colors and sweet scents of woodland wildflowers are advertisements, attracting bees, butterflies, flies, and even beetles to their nectar-rich blooms. I remember once watching a bumblebee meticulously work its way through a patch of bluebells, its fuzzy body coated in pollen. As it moved from flower to flower, it unknowingly performed the essential task of pollination, ensuring the next generation of bluebells.

But it's not just insects that play a role in woodland pollination. Some plants have evolved to attract birds, particularly hummingbirds, with their vibrant red or orange flowers and abundant nectar. These flowers often have a tubular shape, perfectly suited to the hummingbird's long beak and tongue. Other plants rely on mammals like bats or even small rodents for pollination. These animals are often attracted by the strong, musky scents of the flowers, which are typically open at night. The diversity of pollination strategies in the woodland is a testament to the power of co-evolution, the reciprocal adaptation of two species to each other. As plants have evolved to attract specific pollinators, those pollinators have, in turn, evolved to efficiently exploit the resources offered by the plants. This intricate dance of adaptation has shaped the biodiversity of the woodland, creating a complex web of interactions that sustains the entire ecosystem. And, considering the future, preserving these delicate relationships is essential for maintaining the health and resilience of our woodlands. By understanding and appreciating the symbiotic connections that underpin these ecosystems, we can better protect them for generations to come.

4.1 Mycorrhizal Networks

Beneath the leaf litter, a hidden world pulses with life, a network far more extensive and intricate than the branches reaching for the sky. This is the realm of mycorrhizal networks, nature's internet for plants. These networks aren't simply about nutrient exchange; they are the foundation of woodland community, influencing everything from seedling survival to the resilience of entire stands of trees. As a beginner, understanding these underground connections is crucial to appreciating the complexity and interconnectedness of the woodland ecosystem.

The term "mycorrhiza" literally translates to "fungus-root," a fitting descriptor for the symbiotic relationship at its heart. It is the mutually beneficial association between plant roots and specialized fungi. Plant roots, in their quest for nutrients, are limited by their surface area and the immediate vicinity of the soil. That is where fungi come in, these fungi extend their hyphae, thread-like filaments, far beyond the root zone, acting as an extension of the plant's root system, like a giant net. These hyphae are incredibly thin, much finer than plant roots, enabling them to access nutrients and water from tiny pores in the soil that roots cannot reach. The fungus then transports these resources back to the plant in exchange for sugars produced during photosynthesis. It's a classic win-win situation; the plant gains access to a wider range of nutrients and water, and the fungus receives a reliable source of energy.

This relationship isn't always a simple trade. Consider different types of mycorrhizae, each with a unique way of interacting with plant roots. Ectomycorrhizae (ECM) are the most common type in woodlands dominated by trees like oaks, beeches, and pines. These fungi form a sheath around the outside of the root and extend hyphae into the spaces between root cells, but they don't penetrate the cells themselves. In contrast, arbuscular mycorrhizae (AM) are found in the roots of many herbaceous plants and some trees. AM fungi penetrate the cells of the root cortex, forming branched structures called arbuscules, which increase the surface area for nutrient exchange. Think of it as the difference between having a delivery service drop packages at your doorstep (ECM) versus bringing them directly into your living room (AM). Understanding the type of mycorrhizal association prevalent in your local woodland can help you understand the nutrient dynamics and plant community composition.

The real magic happens when these individual mycorrhizal associations link together to form networks. Imagine a vast, interconnected web of fungal hyphae spanning meters, even kilometers, beneath the forest floor. This is the Common Mycorrhizal Network (CMN), and it connects individual plants of the same or different species. The ramifications are profound. Through the CMN, plants can share resources like carbon, nitrogen, phosphorus, and water. A large, established tree, for example, can transfer carbon to a shaded seedling struggling to photosynthesize. In other instances, plants can even warn each other of impending threats, such as insect attacks. When a plant is attacked by herbivores, it releases chemical signals that can travel through the CMN to neighboring plants, priming them to defend themselves. This inter-plant communication system is still being explored, but the evidence suggests that CMNs play a crucial role in maintaining the health and stability of woodland ecosystems.

Think about what this means for conservation. Mycorrhizal networks are easily disrupted by soil disturbance, pollution, and unsustainable forestry practices. Clear-cutting, for example, can sever these networks, leaving seedlings without access to the vital resources they need to survive. Similarly, excessive use of fertilizers can reduce the reliance of plants on mycorrhizal associations, weakening the networks over time. When restoring degraded woodlands, it's crucial to consider the role of mycorrhizal fungi. Inoculating seedlings with beneficial fungi before planting can significantly increase their survival rates and accelerate the recovery of the ecosystem. Protecting existing woodlands from disturbance and promoting sustainable forestry practices are essential for maintaining the integrity of these underground networks and the health of the entire woodland community.

Observing mycorrhizal connections in the field can be tricky, given that they are largely hidden beneath the soil. However, you can look for clues. Fungal fruiting bodies, such as mushrooms and truffles, are often associated with mycorrhizal networks. The presence of certain mushroom species can indicate the presence of specific mycorrhizal associations. For example, bright yellow chanterelles are often found in association with oak and beech trees. You can also examine plant roots to look for signs of mycorrhizal colonization, such as a thickened sheath around the root tip or the presence of hyphae extending into the soil. A magnifying glass can be helpful for this. Remember to disturb the soil as little as possible when making these observations.

The study of mycorrhizal networks is a rapidly evolving field, and there is still much we don't understand. However, one thing is clear: these underground connections are essential for the health and resilience of woodland ecosystems. As you continue to explore the world of woodlands, remember to look beneath the surface and appreciate the intricate web of life that sustains it. Understanding mycorrhizal networks is not just about understanding fungi and plants; it's about understanding the interconnectedness of all living things and the importance of protecting these vital connections. It's a journey into a hidden realm where cooperation reigns, and the scent of the sunshine fields is intertwined with the earthy aroma of the forest floor, a testament to the power of symbiotic relationships in shaping the world around us.

4.2 Pollination Strategies

Consider the vibrant tapestry of a woodland, where every thread, every plant, plays a crucial role in the grand design. Pollination, the transfer of pollen from the male part of a flower (the stamen) to the female part (the pistil), enabling fertilization and seed production, is one of the most vital processes in this ecosystem. Plants have evolved a staggering array of strategies to ensure successful pollination, reflecting the diverse challenges and opportunities presented by their specific environments. I want to walk you through some of them.

Many woodland plants rely on animals, particularly insects, for pollination. These plants entice pollinators with a combination of visual and olfactory cues. Think of the bright, showy petals of wildflowers like trilliums, or the delicate, nodding heads of bellflowers. These colors act as beacons, drawing insects from afar. Scent, too, plays a critical role. The sweet fragrance of honeysuckle, or the more subtle, spicy aroma of some orchids, are olfactory lures, signaling the presence of nectar and pollen rewards. What I find particularly fascinating is the specificity of some of these relationships. Certain plants have co-evolved with specific pollinators, developing highly specialized floral structures and attractants. For example, some orchids mimic the appearance of female insects, attracting male insects that attempt to mate with the flower, inadvertently transferring pollen in the process. This kind of deception is a testament to the evolutionary pressures that have shaped these interactions. Others, rather than deceive, provide very specific rewards. Some plants produce specialized oils that certain bee species collect to provision their nests.

While insect pollination is common, other animals also play a role. Birds, particularly hummingbirds in some regions, are attracted to brightly colored, tubular flowers with copious amounts of nectar. As they probe the flowers for nectar, pollen adheres to their feathers, which they then transfer to other flowers. Mammals, such as bats and even small rodents, can also be pollinators, especially in woodlands with limited light or a high density of nocturnal animals. These animals often pollinate flowers with a strong, musky odor, which is more easily detected in the dark. I never considered rodents as pollinators, but it's a thing!

Not all woodland plants rely on animals for pollination. Wind pollination is another important strategy, particularly for trees and grasses. These plants typically produce large quantities of lightweight pollen that is easily dispersed by the wind. Think of the clouds of pollen released by pine trees in the spring, or the feathery stigmas of grasses that efficiently capture windborne pollen. Wind-pollinated plants often have inconspicuous flowers, as they don't need to attract pollinators with bright colors or fragrances. Instead, they focus on producing large amounts of pollen to increase the chances of successful pollination. This can be a wasteful strategy, as much of the pollen is lost to the wind, but it is effective in open woodlands where wind is readily available.

Beyond these basic strategies, many woodland plants employ more complex pollination mechanisms to maximize their reproductive success. Some plants have evolved self-pollination mechanisms, ensuring pollination even in the absence of pollinators. Others have developed mechanisms to prevent self-pollination, promoting genetic diversity and reducing the risk of inbreeding. These mechanisms can include physical barriers that prevent pollen from reaching the pistil, or chemical signals that inhibit pollen germination on the same flower. Still others have developed both methods. What I think is cool is how that allows the plant to have both methods of reproduction.

The effectiveness of different pollination strategies can be influenced by a variety of environmental factors. Weather conditions, such as wind, rain, and temperature, can affect pollinator activity and pollen dispersal. Habitat fragmentation can reduce pollinator populations, making it more difficult for plants to find mates. Climate change can alter the timing of flowering and pollinator emergence, disrupting the delicate synchrony between plants and their pollinators. I wonder what will happen as the climate continues to change?

Understanding pollination strategies is essential for understanding the overall health and resilience of woodland ecosystems. Pollination is a critical process that supports plant reproduction, which in turn supports the entire food web. By understanding the different ways that plants are pollinated, we can better protect and manage these valuable ecosystems.

Protecting woodland pollinators is crucial for maintaining the health of these ecosystems. This can involve a variety of strategies, such as creating pollinator-friendly habitats, reducing pesticide use, and controlling invasive species. Creating pollinator-friendly habitats can involve planting native wildflowers that provide food and shelter for pollinators, or creating nesting sites for bees and other insects. Reducing pesticide use is essential for protecting pollinators from harmful chemicals. Controlling invasive species can help to restore native plant communities and provide more resources for pollinators.

Woodland plants have evolved diverse and fascinating pollination strategies to ensure their reproductive success. These strategies reflect the complex interactions between plants, pollinators, and the environment. By understanding these interactions, we can better appreciate the intricate beauty and ecological importance of woodland ecosystems. Ultimately, it's a delicate balance that requires our attention and care to ensure these ecosystems thrive for generations to come. We should all pitch in to make sure these pollinators are protected.

5 Chapter 5: Seeds and Spores

The journey of a plant from a mere seed or spore to a thriving organism is a testament to nature's ingenuity. Dispersal, the act of spreading seeds and spores away from the parent plant, is a crucial stage that influences plant distribution, population dynamics, and community structure within woodlands. Understanding the different methods plants employ to achieve this dispersal is essential for appreciating the complexities of woodland ecosystems.

Wind dispersal, or anemochory, is a common strategy in woodlands. Lightweight seeds with specialized structures that act as sails or wings are carried away by the breeze. Consider the dandelion, though not strictly a woodland plant, its seed head explodes with feathery pappus, each carrying a single seed aloft. Similarly, maple trees produce samaras, winged seeds that twirl through the air like miniature helicopters, traveling considerable distances in even a gentle breeze. The effectiveness of wind dispersal depends on several factors: wind speed and direction, seed size and shape, and the height of the parent plant. Taller trees, like oaks and ashes, release their wind-dispersed seeds from higher in the canopy, allowing them to travel further than seeds released from lower-growing plants in the understory. A common tree that uses wind dispersal for their seeds is the birch. The adaptations for wind dispersal are evident in the seed's structure, showcasing nature's design.

Animal dispersal, or zoochory, is another highly effective strategy, relying on animals to transport seeds. This can occur in two main ways: epizoochory, where seeds are attached to the animal's exterior, or endozoochory, where seeds are consumed and later deposited in feces. Burrs, like those produced by burdock, are excellent examples of epizoochory. Their hooked spines latch onto fur or feathers, allowing seeds to hitchhike long distances. Endozoochory, on the other hand, involves fleshy fruits that are attractive to birds and mammals. When an animal eats the fruit, the seeds pass through its digestive system unharmed and are deposited elsewhere, often with a dose of fertilizer. Berries, such as those produced by dogwood and elderberry, are particularly popular with birds, which play a vital role in their dispersal. Squirrels and jays contribute significantly to the dispersal of acorns and other nuts by burying them as food stores. Many of these caches are never recovered, allowing the forgotten seeds to germinate and grow. Animal dispersal is particularly important for plants in the understory, where wind dispersal is less effective due to limited airflow. It is more complex than it looks.

Water dispersal, or hydrochory, is less common in many woodland environments but plays a crucial role in riparian areas or woodlands near bodies of water. Seeds may be buoyant or protected by waterproof outer layers, allowing them to float downstream to new locations. Alder trees, often found near streams and rivers, produce seeds with air-filled sacs that aid in flotation. Water currents can carry these seeds over considerable distances, enabling colonization of new habitats along waterways. This strategy is particularly advantageous in disturbed areas where bare soil is exposed and competition from other plants is reduced. Understanding the characteristics of the seed that allow it to be dispersed in water can tell us the types of plants that can grow in that environment.

Seed germination is the crucial process by which a seed sprouts and begins to grow into a seedling. Germination success depends on a combination of environmental factors and inherent seed characteristics. Adequate moisture is essential for triggering germination, as water is needed to rehydrate the seed and initiate metabolic processes. Temperature is also critical; most seeds have an optimal temperature range for germination, which varies depending on the species. Light availability can also influence germination, with some seeds requiring light exposure to germinate, while others prefer darkness. Seed dormancy, a state of suspended animation, is another important factor that can influence germination success. Many woodland seeds require a period of cold stratification, exposure to cold temperatures over the winter, to break dormancy and prepare them for germination in the spring. Scarification, the process of weakening or breaking the seed coat, can also be necessary for germination, particularly for seeds with hard outer layers. This can occur through physical abrasion, such as passing through an animal's digestive system, or through chemical processes in the soil.

Spores, unlike seeds, are single-celled reproductive units that are dispersed by ferns, mosses, and other non-seed plants. Spore release mechanisms are adapted to ensure effective dispersal. Ferns, for example, have sporangia, specialized structures on the underside of their fronds that release spores when they dry out. The sporangia catapult the spores into the air, where they can be carried by wind currents. Mosses, on the other hand, release spores from capsules located on the tips of stalks. These capsules often have intricate mechanisms for spore dispersal, such as teeth that open and close in response to humidity changes. Spore germination requires specific environmental conditions, including adequate moisture and a suitable substrate. Spores typically germinate into a small, heart-shaped structure called a prothallus, which is the gametophyte generation of the plant. The prothallus produces both sperm and eggs, which fuse to form a zygote that develops into the sporophyte generation, the familiar fern or moss plant. Spore dispersal is essential for the colonization of new habitats and the maintenance of genetic diversity in these plant groups. Without effective dispersal, plant populations may become isolated and vulnerable to local extinction.

In the intricate tapestry of woodland ecosystems, seeds and spores are the threads that ensure the continuation of plant life. By understanding the mechanisms of seed dispersal, germination success, and spore release, we gain a deeper appreciation for the resilience and adaptability of these vital components of our natural world. The survival of plants depends on the success of seeds and spores. Protecting our woodlands also means preserving the processes that help plants to thrive and grow.

5.1 Seed Dispersal Mechanisms

The journey of a seed, from its parent plant to a suitable germination site, is a fascinating tale of adaptation and strategy. For many plants in woodland environments, successful seed dispersal is paramount to their survival. With limited space and fierce competition for resources, plants must rely on creative methods to distribute their offspring far and wide. Understanding these mechanisms is key to appreciating the dynamics of plant populations and the overall health of the ecosystem.

Consider the whispering wind, a tireless worker in the realm of seed dispersal. Some seeds have evolved specialized structures to harness its power. Take the dandelion, a common yet remarkable example. Its seed is attached to a feathery pappus, acting as a miniature parachute. This allows the seed to float gracefully on the breeze, potentially traveling great distances before landing. Other plants, such as maple trees, employ a different strategy. Their seeds are encased in winged fruits, known as samaras, which spin like helicopter blades as they fall, increasing their airtime and dispersal range.

For wind dispersal to be effective, however, the seeds must be lightweight and produced in abundance. This is because wind is an indiscriminate carrier, and most wind-dispersed seeds will not land in a favorable location. However, for those that do, the potential for colonization of new areas is significant.

Water, another vital force, also plays a significant role in seed dispersal, particularly in woodlands near rivers, streams, or wetlands. Plants that rely on water dispersal, known as hydrochory, produce seeds that are buoyant and resistant to water damage. Some seeds possess air-filled cavities or waterproof coatings that allow them to float for extended periods. These seeds can travel long distances downstream, colonizing new areas along the waterway. The water lily, with its floating seeds, is a prime example of a plant that utilizes this method. The effectiveness of water dispersal depends on the speed and direction of the current, as well as the availability of suitable landing sites along the banks.

Perhaps the most captivating seed dispersal method involves animals, a process known as zoochory. Woodland ecosystems are teeming with animals that inadvertently or intentionally carry seeds to new locations. Fleshy fruits are a common lure for birds and mammals. Animals consume the fruits, and the seeds pass through their digestive tracts unharmed. These seeds are then deposited in new locations, often with a dose of natural fertilizer. Berries are a classic example of this strategy, attracting a wide range of birds and mammals. Other plants produce seeds with hooks, barbs, or sticky substances that attach to animal fur or feathers. These seeds hitch a ride on passing animals, eventually falling off in a new location. The burdock plant is a notorious example of this method, its prickly burrs clinging tenaciously to clothing and animal fur alike. Beyond the immediate dispersal, there's a subtle elegance in how some squirrels meticulously bury acorns as winter stores, often forgetting a significant portion, and thus, planting the next generation of mighty oak trees.

Beyond these primary methods, there are other intriguing dispersal strategies. Some plants employ explosive mechanisms to eject their seeds, launching them away from the parent plant. The touch-me-not plant, true to its name, is a prime example. When its seed pods are touched, they burst open, scattering seeds in all directions. This method is particularly effective for plants that grow in dense vegetation, where wind dispersal may be limited. Other plants rely on gravity to disperse their seeds, simply dropping them to the ground below. While this method may not result in long-distance dispersal, it can be effective for colonizing areas directly beneath the parent plant. Additionally, the simple act of rain can splash seeds out of their pods or capsules, creating a localized dispersal effect.

Successful seed dispersal is not simply a matter of getting the seed to a new location. The seed must also land in a suitable environment for germination and establishment. Factors such as soil type, moisture availability, light levels, and competition from other plants all play a role in determining the success of a seedling. Plants have evolved various adaptations to increase their chances of landing in a favorable location. For example, some seeds have specialized structures that help them bury themselves in the soil, while others are coated with chemicals that inhibit the growth of competing plants. This is important, because simply moving a seed is not sufficient, it must also be in a place where it is capable of developing.

The study of seed dispersal is not merely an academic exercise. It has important implications for understanding plant ecology, conservation, and management. By understanding how seeds are dispersed, we can better predict how plant populations will respond to environmental changes, such as climate change or habitat fragmentation. This knowledge can also be used to develop effective strategies for restoring degraded ecosystems or controlling the spread of invasive plant species. As we seek to protect and preserve our woodland ecosystems, a deeper understanding of seed dispersal mechanisms is essential. Consider the interconnectedness of it all; plants rely on external forces, animals, and even their own unique designs to ensure the continuation of their species, forming a delicate balance within the woodland landscape.

I find that understanding how a seed travels and finds its place provides not just insight into plant life, but also a deeper connection to the patterns and rhythms of the natural world. It's a reminder that even the smallest seed can embark on an extraordinary journey, shaping the landscape and contributing to the rich tapestry of life in our woodlands.

5.2 Germination Success

Germination, that magical moment when a seed awakens and begins its journey into a new plant, is far from a guaranteed event. It's a complex process, influenced by a delicate interplay of internal seed characteristics and external environmental factors. For the woodland novice, understanding these factors is crucial to fostering successful plant propagation and contributing to the health and diversity of our precious woodland habitats. Think of it like this: you're not just planting a seed; you're setting the stage for a whole new life.

The first step towards germination success lies in seed selection. Not all seeds are created equal. Viability, the seed's ability to germinate, is paramount. This varies greatly between species and even within a single species, depending on the parent plant's health, growing conditions, and the seed's maturity at harvest. Look for seeds that are plump, firm, and free from any signs of damage or mold. Reputable seed suppliers will often provide information on germination rates, which can give you a good indication of the seed's quality. Another critical factor is seed dormancy. Many woodland plant seeds have evolved dormancy mechanisms to prevent germination during unfavorable conditions, such as winter. This dormancy can be due to a hard seed coat that prevents water uptake, or internal physiological factors within the embryo. Breaking dormancy often requires specific treatments, such as stratification (exposure to cold, moist conditions) or scarification (scratching or weakening the seed coat). Research the specific dormancy requirements of the species you wish to propagate; neglecting this step is a common reason for germination failure. Remember, a seemingly lifeless seed might just be waiting for the right cue.

Now, let's consider the environmental conditions that are essential for germination. Water is the lifeblood of germination. Seeds need to imbibe water to trigger metabolic activity and rehydrate the embryo. However, too much water can be detrimental, leading to fungal infections and oxygen deprivation. The ideal moisture level is consistently damp, but not waterlogged. Think of a wrung-out sponge. Temperature is another critical factor. Each species has an optimal temperature range for germination. Woodland plants, adapted to cooler climates, generally prefer temperatures between 15-25°C (59-77°F). Soil temperature is particularly important, as it directly affects the seed's metabolic rate. A soil thermometer can be a useful tool for monitoring temperature. Finally, light, while not always essential for germination, can play a role for some species. Some seeds require light to germinate, while others are inhibited by it. This is often related to the seed's size and the depth at which it's buried. Smaller seeds, which have limited energy reserves, often need light to stimulate germination near the soil surface.

The planting medium also plays a vital role. Woodland plants thrive in well-drained, slightly acidic soil. A mixture of peat moss, perlite, and vermiculite can provide a good balance of moisture retention, aeration, and drainage. Avoid using garden soil, as it can be too heavy and compacted, hindering root development and increasing the risk of disease. The depth at which you sow your seeds is another crucial consideration. As a general rule, plant seeds at a depth equal to one to two times their diameter. Smaller seeds should be sown closer to the surface, while larger seeds can be planted slightly deeper. After sowing, gently water the soil and cover the container with a clear plastic lid or bag to maintain humidity. Place the container in a warm, well-lit location, but avoid direct sunlight, which can overheat the soil. Regularly monitor the soil moisture and water as needed to keep it consistently damp.

Patience is key. Germination can take anywhere from a few days to several weeks, depending on the species and environmental conditions. Avoid the temptation to constantly check the seeds, as this can disturb them and disrupt the germination process. Once seedlings emerge, gradually acclimate them to outdoor conditions by exposing them to increasing amounts of sunlight and fresh air. This process, known as hardening off, helps them to develop stronger stems and leaves, making them more resistant to environmental stresses. Finally, transplant the seedlings into individual pots or directly into the woodland environment, ensuring that they have adequate space to grow and thrive. Remember to choose a location that provides the appropriate light, moisture, and soil conditions for the species you are planting.

Beyond the basics, consider the power of observation. Keep a detailed record of your germination experiments, noting the species, seed source, dormancy treatments, environmental conditions, and germination rates. This will help you to identify what works best for different species and refine your techniques over time. Don't be discouraged by failures; every unsuccessful attempt is a learning opportunity. By understanding the intricate factors that influence germination, you can increase your chances of success and contribute to the restoration and enhancement of our woodland ecosystems. Your dedication to these tiny seeds will bring forth a beautiful and flourishing world.

5.3 Spore Release

Spores, unlike seeds, are single-celled reproductive units, and their release is a critical event in the life cycle of fungi, ferns, mosses, and other non-seed-bearing plants within woodland environments. Understanding how these spores are liberated into the environment, and the factors influencing their dispersal, is fundamental to comprehending the distribution, abundance, and overall health of these essential woodland inhabitants. The elegance and ingenuity displayed in these processes, often unseen, underscores the hidden complexity of our forests.

The mechanisms of spore release are remarkably diverse, tailored to the specific morphology and ecology of each organism. In fungi, for instance, spore release is often an active process, relying on specialized structures and physiological processes to propel spores into the air. Consider the Pilobolus fungus, a common inhabitant of herbivore dung. This fungus possesses a phototropic sporangium, a structure that bends towards light. As the sporangium matures, pressure builds up inside until it explosively discharges, launching the spore-containing vesicle up to several meters away, increasing the chance that the vesicle will land on vegetation to be consumed by the next herbivore. This is not passive liberation; it is an active ejection, beautifully adapted for spore dispersal. Other fungi utilize different strategies. Some rely on wind dispersal, releasing their spores from elevated structures, such as mushrooms or puffballs, maximizing their exposure to air currents. The puffball, for instance, releases clouds of spores when disturbed by raindrops or passing animals. The physical impact creates a puffing effect that disperses the spores into the air. Still others depend on water droplets to dislodge spores, which are then carried away by the splashing water, a mechanism common in aquatic fungi and those inhabiting damp environments. I recall vividly observing this process in a ravine during a particularly wet spring; the sheer abundance of spores released with each raindrop was astounding, a testament to the power of even the smallest interaction in the natural world.

Ferns, too, have their own fascinating spore release strategies. Typically, fern spores are produced within structures called sporangia, which are often clustered together in sori on the underside of fern fronds. As the sporangia mature, they dry out, causing a specialized structure called the annulus to contract. This contraction eventually creates tension, causing the sporangium to rupture and catapult the spores into the air. The annulus acts as a tiny, incredibly precise catapult, projecting spores with remarkable force. The effectiveness of this mechanism depends heavily on environmental conditions, particularly humidity. In dry conditions, the annulus functions most efficiently, allowing for optimal spore dispersal. I have noticed that fern spore release is particularly noticeable on warm, breezy afternoons, when the conditions are just right for this catapult mechanism to work its magic.

Mosses, while seemingly simple organisms, also possess intriguing mechanisms for spore release. Moss spores are typically contained within capsules atop long stalks called setae. The capsules often have a lid, or operculum, that falls off when the spores are mature. Some mosses have peristome teeth around the opening of the capsule. These teeth are hygroscopic, meaning they change shape in response to changes in humidity. In dry conditions, the teeth open, allowing for spore release, while in wet conditions, they close, preventing the spores from being washed away prematurely. It's a surprisingly intricate system for such a small plant, showcasing the diversity of solutions found in nature to achieve the same goal: dispersal.

The success of spore release is greatly influenced by environmental factors. Wind speed, humidity, temperature, and even the presence of rainfall can all play significant roles. Wind is perhaps the most obvious factor, as it directly influences the distance spores can travel. However, humidity is equally important. Many spore release mechanisms are triggered by changes in humidity, as seen in ferns and mosses. Temperature can also affect spore viability and germination rates, influencing the overall success of spore dispersal. Rainfall, while sometimes aiding in spore dispersal through splash mechanisms, can also hinder it by washing spores away or creating conditions unfavorable for their release. I have seen, for instance, that prolonged periods of heavy rain can actually suppress spore release in many fungi, leading to reduced dispersal and, potentially, localized population declines.

Beyond environmental conditions, the physical characteristics of the spores themselves play a crucial role in their dispersal. Spore size, shape, and surface ornamentation can all influence how far a spore travels and where it eventually lands. Smaller spores are generally more easily dispersed by wind, while larger spores may be more effectively dispersed by animals or water. Some spores have elaborate surface structures, such as spines or hooks, that aid in attachment to surfaces, facilitating dispersal by animals. Others may have hydrophobic coatings that help them float on water, enhancing their dispersal in aquatic environments. The spores released from fungi, for instance, often have diverse ornamentation patterns when looked under a microscope that allow researchers to differentiate types and their means of transport.

The study of spore release is not just an academic exercise; it has practical implications for understanding ecosystem dynamics and conservation efforts. By understanding how spores are dispersed, we can better predict the spread of fungal diseases, the colonization of new habitats by ferns and mosses, and the overall resilience of woodland ecosystems. For example, knowing the dispersal mechanisms of invasive fungal pathogens can help us develop strategies to prevent their spread and protect vulnerable plant species. Similarly, understanding the factors that promote spore germination and establishment can aid in restoration efforts, allowing us to more effectively re-establish native plant communities in degraded areas. It's about connecting the dots, understanding how these microscopic events influence the larger picture of woodland health and biodiversity.

Finally, consider the artistic beauty inherent in spore release. The seemingly invisible clouds of spores, carried by the wind, are a testament to the power and elegance of nature. They represent the potential for new life, the promise of future generations, and the interconnectedness of all living things. Take a moment to appreciate the unseen world around you, the countless spores silently dispersing, shaping the landscape in ways we are only beginning to understand. It is in these small details that the true wonder of the woodland ecosystem resides, a reminder of the intricate processes that sustain life and the importance of preserving these precious environments for future generations.

6 Chapter 6: Woodland Wildlife

The buzzing of a bee, diligently collecting nectar, is a sound synonymous with woodlands during the warmer months. Insects form the base of a complex food web, driving pollination, decomposition, and nutrient cycling. Consider the leaf litter on the forest floor, a haven for countless invertebrates: beetles, springtails, mites, and the larvae of various flies and moths. These creatures are detritivores, feeding on decaying organic matter and breaking it down into simpler compounds, enriching the soil and making it available for plant uptake. This process is vital for maintaining the health and productivity of the woodland. Caterpillars munching on leaves might seem destructive, but they also play a crucial role, providing food for birds and other insectivores, and their frass (droppings) acts as fertilizer. The intricate relationships between insects and plants are often highly specialized. For example, the larvae of certain butterfly species may only feed on the leaves of a specific tree or shrub, making them particularly vulnerable to habitat loss or the decline of their host plant.

Woodlands are alive with birdsong, a symphony of calls and melodies that varies with the seasons and the time of day. These avian inhabitants are far more than just beautiful singers; they are key players in the ecosystem. Many bird species, such as robins, thrushes, and blackbirds, are insectivores, feeding on insects and helping to control their populations. They also consume fruit and berries, playing a vital role in seed dispersal. Imagine a robin feasting on the bright red berries of a hawthorn tree. As it flies from tree to tree, it deposits the undigested seeds in its droppings, effectively planting new hawthorn saplings. Similarly, jays are known for caching acorns, burying them in the ground for later consumption. Many of these acorns are never recovered, and they germinate into oak trees, contributing to the regeneration of the woodland. Birds of prey, such as owls and hawks, occupy the top of the avian food chain, preying on rodents and other small mammals, helping to regulate their populations. The presence and diversity of bird species are often indicators of the overall health of a woodland ecosystem. Changes in bird populations can signal environmental problems such as habitat degradation or pesticide contamination.

The rustling in the undergrowth might reveal a glimpse of a mammalian resident, from tiny voles and mice to larger deer and foxes. Each mammal plays a unique role in shaping the woodland environment. Small rodents are important seed predators, consuming seeds and nuts but also inadvertently aiding in seed dispersal by burying them and forgetting their caches. They also serve as a food source for predators such as owls, foxes, and weasels.

Larger herbivores, such as deer, browse on leaves, twigs, and bark, influencing the composition and structure of the vegetation. Overgrazing by deer can prevent the regeneration of certain tree species and alter the balance of the understory vegetation. Carnivores, such as foxes and badgers, prey on rodents and other small mammals, helping to control their populations and maintain ecosystem stability. They also scavenge on carrion, playing a role in decomposition. The presence of mammals can have a significant impact on the woodland landscape. For example, the burrowing activities of rabbits and badgers can create disturbances in the soil, which can benefit certain plant species and create habitat for other animals.

To attract the inhabitants of the woods to your garden, creating suitable habitats is essential. If you want to attract insects, native wildflowers and shrubs can be planted to provide food and shelter. The avoidance of pesticides is crucial for protecting beneficial insects, which are a natural form of pest control. For birds, providing birdhouses, feeders, and a source of water can be a magnet for the wildlife. Native trees and shrubs can offer nesting sites and food sources. For mammals, providing safe and undisturbed areas with plenty of cover will entice them to visit. Piles of logs or stones will create havens for small mammals and amphibians. Remember that these creatures are just surviving and trying to live in harmony with the environment surrounding them.

Understanding the intricate relationships between woodland wildlife and their environment is essential for effective conservation. Habitat loss and fragmentation are major threats to woodland ecosystems, reducing the availability of food, shelter, and breeding sites for many species. Pollution can also have a detrimental impact, contaminating water sources and harming wildlife. Climate change is altering the distribution and abundance of many species, as well as increasing the risk of extreme weather events such as droughts and floods. To protect woodland wildlife, it is crucial to conserve and restore woodland habitats, reduce pollution, and mitigate the impacts of climate change. This can involve initiatives such as planting trees, creating wildlife corridors, and implementing sustainable forestry practices. By understanding and appreciating the vital role that wildlife plays in woodland ecosystems, we can work together to ensure their long-term health and resilience. These factors provide a better understanding on how to conserve and protect them.

6.1 Insects and Herbivores

Insects and herbivores play an indispensable role in shaping woodland plant communities. Their influence extends far beyond simple consumption, impacting plant distribution, abundance, and even evolutionary trajectories. Understanding these interactions is crucial for comprehending the complex web of life within these ecosystems. From the smallest leaf-miner to the largest deer, herbivores exert selective pressures that drive adaptation and diversification in the plant kingdom.

Herbivory, the act of animals feeding on plants, isn't a simple predator-prey relationship. It's a dynamic interaction, a constant dance of defense and counter-defense. Plants have evolved an arsenal of strategies to deter herbivores, ranging from physical barriers like thorns and spines to chemical defenses such as toxins and repellents. These defenses aren't always static; many plants can induce defenses in response to herbivore attack, becoming less palatable or even toxic when grazed upon. Conversely, herbivores evolve mechanisms to overcome these defenses, leading to an evolutionary arms race that shapes both plant and animal communities. The types of herbivore present in a woodland, and their feeding preferences, can dramatically alter the composition of plant species. For instance, a high deer population can suppress the regeneration of palatable tree species, favoring those with greater browse-resistance. Similarly, insect herbivory can influence the competitive balance between plant species, creating opportunities for less dominant plants to thrive.

Insects constitute the most diverse group of herbivores in woodland ecosystems. From leaf-chewing caterpillars to sap-sucking aphids and wood-boring beetles, their feeding habits impact virtually every part of a plant. Different insect species specialize on different plant parts or plant species, creating intricate feeding webs. For example, certain butterfly larvae may feed exclusively on specific host plants, while other insects are generalists, feeding on a wide range of vegetation. The impact of insect herbivory can range from minor defoliation to complete plant mortality, depending on the insect species, population size, and the plant's vulnerability. Outbreaks of certain insect pests, such as gypsy moths or bark beetles, can cause widespread defoliation and tree mortality, significantly altering forest structure and composition. Even seemingly minor insect feeding can have cascading effects, impacting plant growth, reproduction, and susceptibility to other stresses.

Larger herbivores, such as deer, rabbits, and voles, also play a significant role in shaping woodland plant communities. Their grazing and browsing habits can significantly impact plant biomass, species composition, and regeneration patterns. Deer, for example, are selective browsers, often preferring certain palatable tree seedlings and shrubs over others. This selective browsing can alter the competitive balance among plant species, leading to changes in forest structure and composition. High deer populations can also suppress the regeneration of desirable tree species, impacting timber production and forest biodiversity. Rabbits and voles, while smaller in size, can also exert considerable grazing pressure, particularly on herbaceous plants and seedlings. Their feeding habits can influence ground cover vegetation and the establishment of new tree seedlings. The presence or absence of these herbivores can dramatically alter the understory vegetation and overall diversity of the woodland.

The impact of herbivores on plant communities is not always negative. In some cases, herbivory can promote plant diversity and resilience. For instance, moderate grazing can reduce competitive dominance by certain plant species, creating opportunities for less competitive species to thrive.

Herbivory can also stimulate plant growth and reproduction in some cases. For example, defoliation can sometimes stimulate increased branching and leaf production, while seed dispersal by herbivores can help plants colonize new areas. Furthermore, the presence of herbivores can support other species in the ecosystem. Insects, for instance, serve as a food source for birds, amphibians, and other predators. Larger herbivores can create habitat for smaller animals by opening up dense vegetation. Therefore, the relationship between herbivores and plants is complex and multifaceted, with both positive and negative impacts on the ecosystem.

Protecting woodland plant communities requires a careful understanding of the role of insects and herbivores. Effective management strategies must consider the complex interactions between plants, herbivores, and other ecosystem components. Overpopulation of certain herbivores, such as deer, can be addressed through hunting regulations, habitat management, and predator reintroduction. Insect pest outbreaks can be managed through integrated pest management strategies that combine biological control, cultural practices, and, as a last resort, selective pesticide applications. Preserving a diversity of plant species and habitats is essential for maintaining healthy and resilient woodland ecosystems. By promoting a balanced ecosystem with a diverse array of herbivores and predators, we can ensure the long-term health and vitality of these valuable natural resources. The delicate balance between plant life and animal influence must be taken into consideration in order to maintain a healthy, prosperous ecosystem. After all, the scent of sunshine fields depends on every tiny interaction of life.

6.2 Birds of the Forest

Woodland birds are not just pretty faces flitting through the branches; they're integral players in the delicate balance of the ecosystem. Their presence, or absence, speaks volumes about the health of the woods themselves. We're talking about more than just a charming chorus in the morning – we're talking about seed dispersal, insect control, and even nutrient cycling. So grab your binoculars, because we're diving into the avian world of the woods.

Let's start with the seed spreaders. Think about a majestic oak, its acorns destined to sprout new life. How do those acorns get far enough from the parent tree to avoid competing for resources? Enter the jay, a master of acorn relocation. These clever birds bury acorns for later consumption, often forgetting some in the process. Those forgotten acorns, blessed with a patch of soil and a little luck, become the next generation of oaks. Similarly, berry-eating birds like thrushes and waxwings ingest fruits and deposit the seeds elsewhere through their droppings, distributing plants far and wide. Each bird species, with its preferred diet and foraging habits, contributes to a unique pattern of plant dispersal, shaping the composition and structure of the woodland. Without them, the forest would look very different, with plant life clustered around parent trees and a slower rate of regeneration. These birds play an especially crucial role in post-disturbance recovery;

consider areas where fire or storms have cleared patches, they act as the architects of restoration, ensuring the woodland regrows.

Moving on to insect control, the woodland is a buzzing metropolis of insects. Some are beneficial, of course, but others can wreak havoc on trees and other plants. That's where insectivorous birds come in. Warblers, with their needle-like beaks, glean insects from leaves and branches, controlling populations of leaf-chewing caterpillars and other pests. Woodpeckers, equipped with powerful beaks and long, sticky tongues, excavate insects from beneath the bark, preventing infestations that could weaken or kill trees. Even the seemingly harmless chickadee is a voracious insect eater, consuming vast quantities of aphids and other small pests. These birds provide a natural form of pest control, reducing the need for chemical interventions that can harm the ecosystem. Think of the ecological implications for forest management strategies; by supporting populations of insectivorous birds through habitat protection and the provision of nesting sites, we can reduce the reliance on synthetic pesticides and foster a more resilient and healthy forest.

Consider the intricate web of dependencies within the woodland. Birds are not just consumers; they're also prey. Hawks and owls patrol the skies, keeping populations of smaller birds and rodents in check. These predators play a crucial role in maintaining biodiversity, preventing any one species from becoming dominant and disrupting the balance of the ecosystem. Even the seemingly insignificant songbird contributes to this web, providing a food source for larger predators and indirectly influencing the populations of their prey. This predator-prey dynamic is a driving force in natural selection, shaping the behavior and characteristics of both predators and prey and contributing to the overall resilience of the woodland ecosystem. Understanding these complex interactions is essential for effective conservation management.

The presence and diversity of birds can also serve as an indicator of woodland health. Certain species are particularly sensitive to changes in habitat, pollution, or climate. Their decline or disappearance can be a warning sign that something is amiss. For example, the presence of healthy populations of ground-nesting birds indicates suitable habitat with minimal disturbance. The abundance of insectivorous birds suggests a healthy insect population, which is in turn dependent on a diverse and thriving plant community. By monitoring bird populations, we can gain valuable insights into the overall health of the woodland and identify potential threats before they become catastrophic. This is where citizen science initiatives can make a real impact, as volunteers can help collect data on bird populations and distribution, providing valuable information for researchers and conservation managers.

So, what can we do to protect and enhance bird populations in woodlands? The answer lies in habitat conservation and restoration. Preserving existing woodlands, with their diverse array of tree species, shrubs, and ground cover, is paramount. Creating corridors between fragmented woodlands allows birds to move freely and maintain genetic diversity. Restoring degraded woodlands by planting native trees and shrubs provides food and shelter for birds and other wildlife. In addition, minimizing disturbance from human activities, such as logging and development, is crucial for protecting nesting sites and foraging areas. Promoting sustainable forestry practices that prioritize biodiversity and ecosystem health is essential for ensuring the long-term survival of woodland birds and the ecosystems they inhabit. Furthermore, educating the public about the importance of woodland birds and their role in the ecosystem can foster a sense of stewardship and encourage individuals to take action to protect these valuable resources. Think about planting native trees and shrubs in your backyard, providing bird feeders and bird baths, and reducing your use of pesticides and herbicides. Every little bit helps.

Beyond conservation, understanding the specific needs of different bird species is crucial. Some birds require mature forests with large trees and abundant deadwood, while others thrive in younger, more open woodlands. Providing a variety of habitats within a woodland complex can support a greater diversity of bird species. Furthermore, protecting wetlands and riparian areas within woodlands is essential, as these areas provide critical breeding and foraging habitat for many bird species. Tailoring management practices to the specific needs of different bird species requires a thorough understanding of their ecology and behavior, highlighting the importance of ongoing research and monitoring efforts.

Ultimately, the fate of woodland birds is intertwined with the fate of the woodlands themselves. By protecting and restoring these valuable ecosystems, we not only safeguard the avian inhabitants but also preserve the many other benefits that woodlands provide, from clean air and water to carbon sequestration and recreation. The song of the woodland bird is not just a pleasant sound; it's a reminder of the interconnectedness of life and the importance of protecting our natural heritage. And it is our responsibility to ensure that those songs continue to echo through the trees for generations to come.

6.3 Mammalian Residents

The rustling leaves often conceal more than meets the eye; beneath the canopy's embrace, a vibrant community of mammals thrives, each playing a vital role in the delicate balance of the woodland ecosystem. These creatures, ranging from the diminutive to the substantial, contribute significantly to nutrient cycling, seed dispersal, and overall habitat structure. Consider the humble vole, a seemingly insignificant rodent. Voles, often overlooked, are prolific herbivores, consuming vast quantities of grasses, seeds, and roots. Their constant foraging affects plant growth and distribution, preventing any single species from dominating the understory. Moreover, their tunneling activities aerate the soil, improving drainage and creating pathways for other organisms. They are also, crucially, a vital food source for predators higher up the food chain. This intricate relationship highlights the interconnectedness of even the smallest members of the woodland community. Imagine a world without voles; the delicate balance of the undergrowth would be forever altered, with cascading effects rippling through the entire ecosystem.

Moving up the size scale, we encounter the iconic red squirrel, a creature synonymous with woodland habitats. These agile climbers are primarily seed eaters, playing a pivotal role in seed dispersal. Their habit of caching nuts for later consumption, often burying them in various locations, ensures the regeneration of tree species. Many of these buried caches are never retrieved, resulting in the germination of new seedlings and contributing to the forest's future composition. The red squirrel's presence is a testament to the health and maturity of a woodland, indicating a robust supply of seeds and a complex habitat structure that can support their arboreal lifestyle. Sadly, the introduction of the grey squirrel in some regions has created fierce competition, threatening the native red squirrel population and highlighting the vulnerability of ecosystems to invasive species.

Deer, particularly the white-tailed deer in North America or the roe deer in Europe, are perhaps the most visible mammalian residents of many woodlands. Their browsing habits significantly influence vegetation structure, favoring certain plant species over others. Overpopulation of deer can lead to overgrazing, reducing plant diversity and hindering the regeneration of young trees. However, at sustainable population levels, deer contribute to the creation of diverse habitats by opening up areas for sunlight penetration and promoting the growth of wildflowers and other ground cover plants. Managing deer populations is therefore crucial for maintaining a healthy and balanced woodland ecosystem. Consider the intricate dance between deer and vegetation, a constant push and pull that shapes the landscape. Then there are the predators, the carnivores that keep herbivore populations in check and maintain the overall health of the ecosystem. Foxes, with their cunning and adaptability, are common woodland predators, preying on rodents, birds, and insects. Their presence helps regulate populations of smaller mammals, preventing overgrazing and maintaining a balanced food web. Similarly, the elusive wildcat, a denizen of more remote woodlands, plays a similar role, preying on a variety of small animals. These predators are essential for preventing any single species from becoming overly dominant and disrupting the ecosystem's delicate equilibrium. The silent hunt, the unseen chase – these are vital processes that maintain the health and diversity of the woodland.

Badgers, often overlooked despite their substantial size, are ecosystem engineers, creating extensive burrow systems that provide shelter for themselves and other animals. These burrows aerate the soil and provide nesting sites for various species, contributing to habitat complexity and biodiversity. Badgers are also opportunistic omnivores, consuming a wide range of foods, including insects, earthworms, and fruits. Their diverse diet further contributes to the intricate web of interactions within the woodland ecosystem. Their sett-building activities physically reshape the environment, creating microhabitats that benefit a multitude of other species.

Smaller mammals, such as shrews and hedgehogs, also play important roles. Shrews, with their voracious appetites, consume vast quantities of insects, helping to control populations of potential pests. Hedgehogs, with their spiny defenses, are also insectivores, contributing to the regulation of insect populations. These seemingly insignificant creatures play a vital role in maintaining the balance of the woodland ecosystem, preventing outbreaks of pests and ensuring the health of the plant communities. It is easy to overlook these small creatures, but their combined impact on the ecosystem is substantial.

Finally, consider the bats, the nocturnal denizens of the woodland. These flying mammals are primarily insectivores, consuming vast quantities of insects each night, including moths, mosquitoes, and beetles. Their role in controlling insect populations is invaluable, reducing the need for pesticides and ensuring the health of the forest. Bats also play a role in pollination and seed dispersal, particularly in tropical woodlands. Their silent flight and nocturnal habits make them mysterious and often misunderstood, but their ecological importance is undeniable.

The mammalian residents of the woodland are not simply isolated creatures; they are integral components of a complex and interconnected ecosystem. Their actions, from foraging and burrowing to predation and seed dispersal, shape the landscape and influence the lives of countless other organisms. Understanding their roles and their interactions is crucial for effective woodland conservation and management. Protecting these animals and their habitats is essential for maintaining the health and resilience of these valuable ecosystems. Without their presence, the "Scent of Sunshine Fields" would be a fainter, less vibrant fragrance indeed. Each mammal, a thread in the rich tapestry of the woodland, contributing to its beauty, its diversity, and its enduring vitality. They are not just inhabitants; they are essential architects of this complex and fascinating world.

7 Chapter 7: Conservation & Future

Woodlands, these intricate tapestries of life, face a barrage of threats in the modern world, each capable of unraveling the delicate balance that sustains them. Habitat loss, driven by urbanization, agriculture, and unsustainable logging practices, stands as a primary culprit. As forests are fragmented and converted into other land uses, the intricate web of relationships between plants, animals, and microorganisms is disrupted. Imagine a once-continuous expanse of ancient woodland, now bisected by roads and fragmented into isolated patches. This fragmentation hinders the movement of wildlife, reduces genetic diversity, and increases the vulnerability of woodland species to local extinction. Moreover, pollution, in its various forms, casts a long shadow over woodland ecosystems. Air pollution, resulting from industrial emissions and vehicle exhaust, can acidify soils, damage tree foliage, and impair plant growth. Water pollution, stemming from agricultural runoff and industrial discharge, contaminates streams and wetlands, threatening aquatic life and impacting the overall health of the ecosystem. Furthermore, the insidious creep of climate change presents an unprecedented challenge to woodlands worldwide. Rising temperatures, altered precipitation patterns, and increased frequency of extreme weather events are already impacting woodland ecosystems, shifting species distributions, increasing the risk of wildfires, and exacerbating the spread of invasive species. Consider the iconic oak woodlands of California, increasingly vulnerable to drought and wildfires, or the boreal forests of Canada and Russia, threatened by thawing permafrost and altered growing seasons. These complex and interacting threats demand concerted conservation efforts, spanning local, national, and international scales. Protected areas, such as national parks, nature reserves, and wilderness areas, play a vital role in safeguarding woodland habitats from development and exploitation. These protected areas serve as refuges for biodiversity, providing essential habitat for threatened and endangered species. However, simply designating protected areas is not enough. Effective management is crucial to ensure that these areas are adequately protected from illegal logging, poaching, and other destructive activities. Restoration projects, aimed at repairing degraded woodland ecosystems, are also essential. These projects may involve replanting native trees and shrubs, removing invasive species, and restoring hydrological processes. Imagine a degraded woodland, scarred by past logging operations, slowly recovering as native trees are replanted, invasive species are removed, and natural regeneration is encouraged. This restoration process can take decades, but the long-term benefits for biodiversity and ecosystem health are immense. Furthermore, community engagement is vital for successful conservation efforts. Local communities often have a deep understanding of their surrounding environment and can play a crucial role in monitoring woodland health, reporting illegal activities, and participating in restoration projects.

Sustainable practices offer a pathway toward harmonizing human activities with the long-term health of woodland ecosystems. Responsible forestry practices, such as selective logging and reduced-impact harvesting, can minimize the environmental impact of timber extraction while still providing economic benefits. Selective logging involves carefully selecting trees for harvest, leaving the surrounding forest largely intact. Reduced-impact harvesting techniques, such as using smaller machinery and avoiding logging during wet periods, can minimize soil disturbance and protect water quality. In addition, promoting the use of sustainably sourced wood products can reduce the demand for timber from unsustainable sources. Look for certifications like the Forest Stewardship Council (FSC) label, which indicates that wood products come from responsibly managed forests. Ecological restoration practices, focused on restoring degraded ecosystems to a more natural state, are essential for enhancing woodland resilience. These practices may involve reintroducing native species, restoring natural fire regimes, and controlling invasive species. Consider a former agricultural field adjacent to a woodland being restored to native vegetation, providing habitat for wildlife and expanding the overall area of the woodland ecosystem. These restoration efforts can enhance biodiversity, improve water quality, and increase the resilience of woodlands to climate change.

As we look to the future, it is clear that safeguarding woodlands requires a multifaceted approach, combining conservation efforts with sustainable practices and community engagement. Education and awareness are crucial for fostering a greater appreciation for the value of woodland ecosystems and encouraging responsible stewardship. By raising awareness about the threats facing woodlands and promoting sustainable practices, we can inspire individuals, communities, and policymakers to take action. Imagine a school group learning about the importance of woodland ecosystems and participating in a tree-planting project, or a community organizing a volunteer workday to remove invasive species from a local woodland. These hands-on experiences can foster a sense of connection to nature and inspire a lifelong commitment to conservation. Ultimately, the fate of woodlands rests in our hands. By embracing a holistic approach to conservation and sustainable management, we can ensure that these vital ecosystems continue to thrive for generations to come. Let us work together to protect these precious havens of biodiversity and secure a future where the scent of sunshine fields continues to fill the air. We must learn to value woodlands not just for the resources they provide, but for the essential role they play in maintaining the health of our planet and the well-being of all living things. This requires a shift in mindset, from viewing woodlands as commodities to recognizing them as invaluable ecosystems that deserve our respect and protection. It's time to act.

7.1 Threats to Woodlands

Woodlands, for all their beauty and resilience, are not immune to harm. A myriad of threats, both natural and anthropogenic, loom large, jeopardizing the delicate balance of these ecosystems and the countless species that call them home. Comprehending these challenges is paramount to effective conservation and ensuring the longevity of our precious woodlands.

One of the most pervasive and devastating threats is habitat loss. Urban sprawl, agricultural expansion, and infrastructure development relentlessly encroach upon woodland areas, fragmenting once-contiguous habitats into smaller, isolated patches. This fragmentation isolates populations, reducing genetic diversity and limiting the ability of species to migrate, adapt to changing conditions, and find suitable mates. As a result, populations decline, local extinctions occur, and the overall resilience of the woodland ecosystem is compromised. The consequences ripple throughout the food web, impacting everything from apex predators to specialized pollinators.

Pollution, in its many forms, poses a significant threat to woodland health. Atmospheric pollutants, such as acid rain and ozone, damage foliage, impair photosynthesis, and weaken trees, making them more susceptible to disease and insect infestations. Soil pollution from industrial activities, mining, and agricultural runoff contaminates the ground, disrupting nutrient cycles, harming soil microorganisms, and poisoning plants. Water pollution from sewage, fertilizers, and pesticides contaminates streams and wetlands within the woodland, harming aquatic life and impacting the overall water quality of the ecosystem. The insidious effects of pollution can accumulate over time, leading to long-term degradation of woodland habitats.

Climate change is another formidable challenge, exacerbating existing threats and introducing new ones. Rising temperatures, altered precipitation patterns, and increased frequency of extreme weather events, such as droughts, floods, and wildfires, are already impacting woodland ecosystems around the globe. Changes in temperature and precipitation can disrupt plant phenology, affecting flowering times, leaf emergence, and seed dispersal. Shifts in climate can also alter the distribution of species, leading to mismatches in ecological interactions and the displacement of native species by invasive ones. Increased frequency and intensity of wildfires can devastate woodland areas, destroying habitats and releasing vast amounts of carbon dioxide into the atmosphere, further accelerating climate change.

Invasive species, both plant and animal, represent a significant threat to woodland biodiversity. These non-native species, often introduced accidentally or intentionally by humans, can outcompete native plants for resources, alter habitat structure, and disrupt ecological processes. Invasive plants, such as garlic mustard, kudzu, and Japanese honeysuckle, can rapidly spread, forming dense monocultures that displace native vegetation and reduce habitat for wildlife. Invasive animals, such as the emerald ash borer, gypsy moth, and feral pigs, can cause widespread damage to trees, defoliate forests, and disrupt soil ecosystems. The introduction and spread of invasive species can have cascading effects throughout the woodland ecosystem, leading to a decline in biodiversity and a loss of ecosystem services.

Unsustainable forestry practices, such as clear-cutting and high-grading, can also degrade woodland habitats. Clear-cutting removes all trees from a large area, disrupting soil ecosystems, altering water cycles, and eliminating habitat for many species. High-grading involves removing the most valuable trees from a stand, leaving behind less desirable trees and reducing the genetic diversity of the forest. These unsustainable practices can lead to soil erosion, water pollution, loss of biodiversity, and reduced carbon sequestration capacity. Sustainable forestry practices, such as selective harvesting and reforestation with native species, are essential for maintaining the health and resilience of woodland ecosystems.

Overgrazing by livestock can also damage woodlands, particularly in drier regions. Excessive grazing can remove ground cover, compact soil, and prevent tree regeneration, leading to soil erosion, loss of biodiversity, and degradation of habitat. Proper grazing management, such as rotational grazing and fencing to exclude livestock from sensitive areas, is essential for preventing overgrazing and maintaining the health of woodland ecosystems.

Finally, diseases and pests can pose a significant threat to specific tree species and overall woodland health. Introduced pathogens and insect pests, often arriving from other continents, can devastate entire tree populations, altering forest structure and composition. Examples include Dutch elm disease, chestnut blight, and the aforementioned emerald ash borer. Monitoring for these threats, implementing preventative measures, and developing strategies for managing outbreaks are crucial for protecting woodlands from devastating losses.

These threats are interconnected and often act synergistically, amplifying their negative impacts on woodland ecosystems. Addressing these challenges requires a comprehensive and integrated approach, involving government policies, community engagement, scientific research, and sustainable management practices. Only through concerted efforts can we hope to mitigate these threats and ensure the long-term health and resilience of our precious woodlands, preserving their ecological integrity and the many benefits they provide for generations to come. The task is daunting, but the stakes are undeniably high, demanding our immediate attention and unwavering commitment.

7.2 Conservation Efforts

The importance of conservation in woodland ecosystems cannot be overstated. These vital habitats, brimming with biodiversity and providing numerous ecological services, face unprecedented threats in the modern era. From habitat fragmentation and invasive species to the looming crisis of climate change, the challenges are multifaceted and demand a concerted, proactive approach. To truly safeguard these essential environments for future generations, it's essential that we dive into effective conservation strategies.

One of the most effective approaches to conservation revolves around establishing protected areas. National parks, nature reserves, and wilderness areas serve as vital refuges for woodland ecosystems. These designated spaces provide a safe haven for native flora and fauna, allowing natural processes to unfold with minimal human interference. Effective management plans are crucial for these protected areas, ensuring that they truly serve their purpose. This can include regular monitoring of species populations, controlled burns to manage undergrowth and reduce fire risk, and active measures to combat poaching or illegal logging. Establishing buffer zones around protected areas is also key. These zones act as a transitional space, limiting development and intensive agriculture close to the protected core, thereby minimizing the impact of human activities on the sensitive woodland ecosystem within. Consider the example of the Białowieża Forest in Poland and Belarus, one of the last and largest remaining parts of the immense primeval forest that once stretched across the European Plain. Through strict protection measures, including limitations on logging and hunting, this forest has become a crucial stronghold for the European bison and a wealth of other woodland species.

Habitat restoration represents another critical pillar of woodland conservation. Decades of deforestation, unsustainable logging practices, and agricultural expansion have left many woodland ecosystems fragmented and degraded. Active restoration efforts are essential to reconnect these fragmented habitats, improve the health of degraded areas, and enhance the overall resilience of the landscape. Reforestation, planting native tree species, can rebuild woodland areas that have been lost or damaged. It's paramount to use locally sourced seeds and saplings to ensure genetic diversity and adaptation to local conditions. Invasive species are a major threat to woodland ecosystems, outcompeting native plants, disrupting food webs, and altering habitat structure. Removal of these invasive species, often through manual labor, targeted herbicide application, or biological control methods, is a crucial step in restoring the natural balance of the ecosystem. Soil erosion can be a significant problem in degraded woodlands, hindering plant growth and reducing water quality. Implementing soil conservation measures, such as planting cover crops, constructing terraces, and using erosion control blankets, can help stabilize the soil and promote vegetation recovery. The Knepp Estate in West Sussex, UK, exemplifies the power of habitat restoration. By allowing the land to revert to a more natural state through rewilding, the estate has witnessed a remarkable recovery of biodiversity, including rare species of butterflies, birds, and plants, demonstrating the potential of habitat restoration to transform degraded landscapes into thriving woodland ecosystems.

Sustainable forestry practices play a vital role in balancing timber production with the long-term health and integrity of woodland ecosystems.

Traditional logging methods can have devastating impacts, leading to soil erosion, habitat loss, and reduced biodiversity. Sustainable forestry, on the other hand, aims to minimize these negative impacts and ensure that forests are managed in a way that supports both economic and ecological values. Selective logging, where only mature trees are harvested while leaving younger trees and understory vegetation intact, reduces the overall impact on the forest structure and minimizes soil disturbance. Employing reduced-impact logging techniques, such as using lighter machinery and carefully planning logging roads, can further minimize environmental damage. Promoting natural regeneration, allowing forests to regenerate naturally from seeds and seedlings, reduces the need for artificial planting and maintains genetic diversity. Forest Stewardship Council (FSC) certification provides a way for consumers to identify timber products that have been harvested from sustainably managed forests, promoting responsible forestry practices.

Beyond the ecological interventions, public awareness and education are crucial components of any successful conservation strategy. Many people are simply unaware of the importance of woodlands and the threats they face. Raising awareness about the ecological services provided by woodlands, such as clean air and water, carbon sequestration, and habitat for wildlife, can foster a greater appreciation for these ecosystems and inspire people to take action. Educational programs, workshops, and guided tours can provide people with a deeper understanding of woodland ecology and the importance of conservation. Engaging local communities in conservation efforts, such as tree planting events, volunteer monitoring programs, and citizen science projects, can foster a sense of ownership and responsibility for the health of local woodlands. Supporting environmental organizations and conservation groups that are working to protect woodlands is another way to contribute to conservation efforts. These organizations often rely on donations and volunteer support to carry out their work.

Finally, addressing the overarching threat of climate change is essential for the long-term survival of woodland ecosystems. Climate change is already having a profound impact on woodlands, altering temperature and precipitation patterns, increasing the frequency and severity of wildfires, and creating conditions that favor invasive species. Reducing greenhouse gas emissions through transitioning to renewable energy sources, improving energy efficiency, and promoting sustainable transportation is crucial for mitigating the impacts of climate change on woodland ecosystems. Assisting woodlands to adapt to the changing climate, such as through promoting drought-resistant tree species, managing forest density to reduce fire risk, and creating corridors to allow species to migrate, can enhance their resilience. Supporting policies and regulations that promote climate change mitigation and adaptation is essential for protecting woodlands from the worst impacts of climate change.

7.3 Sustainable Practices

Sustainable Practices

The long-term health and vitality of our woodlands depend on the implementation of sustainable practices. These practices extend beyond simply planting new trees; they encompass a holistic approach that considers the intricate web of life within the ecosystem and aims to minimize human impact. It is about recognizing that woodlands are not just collections of trees, but complex, interconnected communities that provide invaluable services, from carbon sequestration to wildlife habitat. Ultimately, understanding and adopting these practices is crucial for ensuring that future generations can experience the wonders of these vital ecosystems.

One of the most crucial sustainable practices is responsible forestry, which is an approach to timber harvesting and forest management that prioritizes the long-term health and resilience of the woodland ecosystem. This goes beyond simply replanting trees after logging; it involves careful planning to minimize soil erosion, protect water quality, and preserve biodiversity. Selective logging, for instance, allows for the removal of specific trees while leaving the majority of the forest intact. This method reduces the overall disturbance to the ecosystem and allows the remaining trees to thrive. Furthermore, responsible forestry emphasizes the importance of maintaining a diversity of tree species and age classes, creating a more resilient and stable forest structure that can withstand environmental changes and disturbances. Avoiding clear-cutting, which involves removing all trees in a given area, is a key aspect of responsible forestry. Clear-cutting can have devastating consequences for woodland ecosystems, leading to soil erosion, habitat loss, and reduced biodiversity. By adopting more sustainable harvesting practices, we can ensure that timber is harvested in a way that minimizes negative impacts on the environment.

Beyond responsible forestry, ecological restoration plays a critical role in revitalizing degraded woodland habitats. This involves actively restoring the natural processes and biodiversity of an ecosystem that has been damaged by human activities or natural disasters. Ecological restoration can take many forms, from removing invasive species to reintroducing native plants and animals. One of the most effective approaches is to focus on restoring the natural hydrology of the woodland, ensuring that water flows properly through the ecosystem. This can involve restoring wetlands, removing dams or other barriers to fish passage, and re-establishing natural stream channels. Another important aspect of ecological restoration is to improve soil health. Healthy soil is essential for plant growth and nutrient cycling. This can involve adding organic matter to the soil, reducing soil compaction, and controlling erosion. It is a long-term process that requires careful planning, monitoring, and adaptation. By actively restoring degraded woodlands, we can enhance their ecological value and resilience.

Managing invasive species is also paramount in maintaining woodland health. Invasive species are non-native plants and animals that can outcompete native species, disrupt ecological processes, and alter habitat structure. They can spread rapidly and cause significant damage to woodland ecosystems, outcompeting native plants for resources like sunlight, water, and nutrients. Effective management of invasive species requires a multifaceted approach that includes prevention, early detection, and control. Preventing the introduction of invasive species is the most effective strategy. This can involve implementing strict quarantine measures, inspecting imported goods, and educating the public about the risks of introducing non-native species. Early detection is crucial for controlling invasive species before they become widespread. This requires regular monitoring of woodland ecosystems for new infestations. Control measures can include manual removal, herbicide application, and biological control. Manual removal involves physically removing invasive plants by hand, which can be effective for small infestations. Herbicide application can be used to control larger infestations, but it is important to use herbicides selectively to avoid harming native plants. Biological control involves using natural enemies, such as insects or pathogens, to control invasive species. It is a complex and challenging task, but it is essential for protecting the biodiversity and ecological integrity of woodland ecosystems.

Conserving soil health is essential for woodland ecosystems. Healthy soil provides a foundation for plant growth, nutrient cycling, and water filtration. Unfortunately, many woodland soils have been degraded by human activities, such as deforestation, agriculture, and urbanization. These activities can lead to soil erosion, compaction, and loss of organic matter, which can significantly reduce soil fertility and water-holding capacity. Soil erosion is a major problem in many woodlands, particularly on steep slopes and areas with sparse vegetation cover. Erosion can lead to the loss of topsoil, which is the most fertile layer of soil. This, in turn, can reduce plant growth and increase sedimentation in streams and rivers. Sustainable forestry practices, such as selective logging and reduced-impact logging techniques, can help minimize soil erosion. Replanting trees and shrubs on eroded areas can also help stabilize the soil and prevent further erosion. Soil compaction is another common problem in woodlands, particularly in areas that have been heavily used by vehicles or livestock. Compaction reduces the pore space in the soil, which can limit water infiltration and root growth. Avoiding heavy machinery and livestock grazing can help prevent soil compaction. In addition, adding organic matter to the soil can help improve soil structure and reduce compaction.

Community involvement and education are fundamental to ensuring the success of sustainable practices in woodlands. Engaging local communities in conservation efforts can foster a sense of ownership and responsibility for the long-term health of the ecosystem. Educational programs can raise awareness about the importance of woodlands, the threats they face, and the actions that individuals and communities can take to protect them. Community involvement can take many forms, from volunteering in restoration projects to participating in citizen science initiatives. Restoration projects provide opportunities for community members to actively contribute to the health of their local woodlands. Citizen science initiatives involve community members in collecting data on woodland health, such as monitoring plant and animal populations, assessing water quality, and tracking invasive species. This data can be used to inform conservation efforts and track the effectiveness of management strategies. Educating the public about the importance of woodlands can inspire them to take action to protect these valuable ecosystems.

www.ingramcontent.com/pod-product-compliance
Lightning Source LLC
LaVergne TN
LVHW022005200926
848481LV00007B/2248